The
Practical
Beekeeper

Volume I
Beginning
Beekeeping Naturally

by Michael Bush

The Practical Beekeeper Volume I
Beginning Beekeeping Naturally

Cover Photo © 2011 Alex Wild www.alexanderwild.com

ISBN: 978-161476-061-0

X-Star Publishing Company
Founded 1961

Dedication

This book is dedicated to Ed and Dee Lusby who were the real pioneers of modern natural beekeeping methods that could succeed with the Varroa mites and all the other new issues. Thank you for sharing it with the rest of us.

About the Book

This book is about how to keep bees in a natural and practical system where they do not require treatments for pests and diseases and only minimal interventions. It is also about simple practical beekeeping. It is about reducing your work. It is not a main-stream beekeeping book. Many of the concepts are contrary to "conventional" beekeeping. The techniques presented here are streamlined through decades of experimentation, adjustments and simplification. The content was written and then refined from responding to questions on bee forums over the years so it is tailored to the questions that beekeepers, new and experienced, have.

In place of an index there is a very detailed Table of Contents.

It is divided into three volumes and this edition contains only one: Volume I, Beginning Beekeeping Naturally.

Acknowledgments

I'm sure I will forget to list many who have helped me along this path. For one thing many were often only known by the names they used on the many bee forums where they shared their experiences. But among those who are still helping, Dee, of course, Dean and Ramona, and all of the wonderful people of the Organic Beekeeping Group on Yahoo. Sam, you are always an inspiration. Toni, Christie thanks for your encouragement. All of you on the forums who asked the same questions over and over, because you showed me what needed to be in this book and motivated me to write down the answers. And of course all of you who insisted I put this in the form of a book.

Foreword

I feel like G.M. Doolittle when he said he had already offered all of what he had to say for free in the bee journals and yet people kept asking for a book. I have virtually all of this on my website and have posted all of it many times on the bee forums. But many people have asked for a book. There is a little new here, and most of it is available for free already on my web site (www.bushfarms.com/bees.htm). But many of us understand the transient nature of the medium of the web and want a solid book on our shelf. I feel the same. So here is the book that you could already have read for free but you can hold it in your hands and put it on the shelf and know you have it.

I've done a lot of presentations and a few have been posted on the web. If you have an interest in hearing some of this presented by me try a web search for videos for "Michael Bush beekeeping" or other topics such as "queen rearing". The material here is also on www.bushfarms.com/bees.htm along with PowerPoint presentations from my speaking engagements.

Table of Contents

Volume I Beginning

BLUF

Learn from the bees

> *"Let the bees tell you"*-Brother
> Adam

BLUF stands for Bottom Line Up Front. That's what this chapter is. I am going to give you the short-cut to success in beekeeping right here and now. Not that the rest isn't worth reading, but the rest is merely elaboration and details. With apologies to C.S. Lewis (who said in *A Horse and His Boy*, *"no one teaches riding quite as well as a horse"*) I think you need to realize that *"no one teaches beekeeping quite as well as bees."* Listen to them and they will teach you.

Trust the Bees

> *"There are a few rules of thumb that are useful guides. One is that when you are confronted with some problem in the apiary and you do not know what to do, then do nothing. Matters are seldom made worse by doing nothing and are often made much worse by inept intervention." —The How-To-Do-It book of Beekeeping, Richard Taylor*

If the question in your mind starts "how do I make the bees ..." then you are already thinking wrong-ly. If your question is "how can I help them with what they are trying to do..." you are on your way to becoming a beekeeper.

Resources

Here, then, is the short answer to every beekeeping issue. *Give them the resources to resolve the problem and let them. If you can't give them the resources, then limit the need for the resources.*
For instance if they are being robbed, what they need is more bees to defend the hive, but if you can't give them that, then reduce the entrance to one bee wide and you will create the "pass at Thermopylae where numbers count for nothing". If they are having wax moth issues in the hive, what they need are more bees to guard the comb. If you can't give them that then reduce the area they need to guard by removing empty combs and empty space.
In other words, give them resources or reduce the need for the resources they don't have.

Panacea

Most bee problems come back to queen issues.

There are few solutions as universal in their application and their success, than adding a frame of open brood from another hive every week for three weeks. It is a virtual panacea for any queen issues. It gives the bees the pheromones to suppress laying workers. It gives them more workers coming in during a period where there is no laying queen. It does not interfere if there is a virgin queen. It gives them the resources to rear a queen. It is virtually foolproof and does not require finding a queen or seeing eggs or accurately diagnosing the problem. If you have any issue with queenrightness, no brood, worried that there is no queen, this is the simple solution that re-

quires no worrying, no waiting, no hoping and no guessing. You just give them what they need to resolve the situation. If you have any doubts about the queenrightness of a hive, give them some open brood and sleep well. Repeat once a week for two more weeks if you still aren't sure. By then things will be well on their way to being fine.

If you are afraid of transferring the queen from the queenright hive, because you are not good at finding queens, then shake or brush all the bees off before you give it to them.

If you are concerned about taking eggs from another new package or small colony, keep in mind that bees have little invested in eggs and the queen can lay far more eggs than a small colony can warm, feed and raise. Taking a frame of eggs from a small struggling new hive and swapping it for an empty comb or any drawn comb will have little impact on the donor colony and may save the recipient if they are indeed queenless. If the recipient didn't need a queen it will fill in the gap while the new queen gets mated and not interfere with things.

It saves a lot of worry and a lot of judging. Instead you can give them the resources and then observing what they do will probably give you a pretty good clue what really was going on. If they don't raise a queen, there is probably a virgin loose in there. If they do raise a queen, they obviously didn't have one or the one they had was not sufficient.

Why this book?

I suppose you'd have to be living under a rock these days to have not heard that the honey bees and beekeepers are in trouble. The problems are complex, far reaching and mostly recent. They are certainly a threat to the survival of the beekeeping industry but, even more so, to the survival of many plants which we need or want for food and many other plants, that are a necessary part of the environment.

> *"People who say it cannot be done should not interrupt those who are doing it."-George Bernard Shaw*

It seems like there is some controversy over whether it is even possible to keep bees without treatments. But there are many of us who are doing this and succeeding.

While most of us beekeepers spend a lot of effort fighting with the Varroa mites, I'm happy to say my biggest problems in beekeeping now are things like trying to get nucs through the winter here in Southeastern Nebraska and coming up with hives that won't hurt my back from lifting or simpler ways to feed the bees.

So my purpose is first of all to talk about how to deal with the current problems of beekeeping, and second of all how to work less and accomplish more at beekeeping.

Let's do a short overview of the problems in beekeeping and the solutions. The details are in the subsequent chapters and volumes.

Unsustainable beekeeping system

Beekeeping Pests

So why are we having problem? We have a lot of recent pests and diseases that have made it to North America (and most other places in the world) in the last 30 years or so. (See the chapter *Enemies of the Bees*) As someone once said, "You can't keep bees like grand-pa did cause grandpa's bees are dead." Most of us beekeepers have lost all of our bees one time or another in the last few decades and this seems to be getting worse. So part of the problem for beekeepers is the pests, but there are other issues.

Shallow Gene Pool

We have a narrow gene pool to start with here in North America and between pesticides, pests, and overzealous programs to control Africanized Honey Bees, many of the pockets of feral bees have been depleted leaving only the queens that people buy. When you consider that there are only a handful of queen breeders providing 99% of the queens, that's a pretty small gene pool. This deficiency used to be made up by feral bees and people rearing their own queens. But the recent trend is to encourage everyone to not rear their own queens and only buy them; especially in AHB (Africanized Honey Bee) areas.

Contamination

The other side of the pest issue is that the standard answer offered by the experts has been to use pesticides in the hives by beekeepers to kill the mites and other pests. But these build up in the wax and cause sterile drones which in turn causes failing queens. One estimate I heard from one of the experts on the subject put the average supersedure rate at three times a year. That means the queens are failing and being replaced three times a year. This is stunning to me since most of my queens are three years old.

Wrong Gene Pool

The other side of helping bees with treatments of pesticides and antibiotics is that you keep propagating the bees that can't survive. This is the opposite of what we need. We beekeepers need to be propagating the ones that *can* survive. Also we keep propagating the pests that are strong enough to survive our treatments. So we keep breeding wimpy bees and super pests. Also for years we have bred bees to not rear drones, be larger, and use less propolis. Some of these make them reproductively challenged (less drones and larger bees hence larger slower drones) and some make them less able to handle viruses (less propolis).

Upset ecology of the bee colony

A bee colony is a whole system in itself of beneficial and benign fungi, bacteria, yeasts, mites, insects and other flora and fauna that depend on the bees for their livelihood and which the bees depend on to ferment the pollen and crowd out pathogens. All of the pest controls tend to kill the mites and insects. All of the antibiotics used by beekeepers tend to kill either the bacteria (Terramycin, Tylosin, essential oils, organic acids and thymol do this) or the fungi and yeasts (Fumidil, essential oils, organic acids and thymol do this). The whole balance of this precarious system has been upset by all the treatments in the hive. And recently beekeepers switched to a new antibiotic, Tylosin, which the beneficial bacteria has not had a chance to build up resistance to and which is longer lived; and they have switched to formic acid as a treatment which shifts the pH radically to the acidic and kills many of the microorganisms of the hive.

Beekeeping House of Cards

So beekeepers, with the advice and assistance of the USDA and the universities, have built this precarious system of beekeeping that relies on chemicals, antibiotics and pesticides to keep it going. And beekeepers keep breeding the resistant pests that can survive the treatments, contaminating the entire wax supply with poisons (and we make our foundation out of that contaminated wax so it is a closed system) and breeding queens that can't survive without all of this treatment.

How do we get a sustainable beekeeping system?

Stop treating

The only way to have a sustainable system of beekeeping is to stop treating. Treating is a death spiral that is now collapsing. To leverage this, though, you really need to raise your own queens from local surviving bees. Only then can you get bees who genetically can survive and parasites that are in tune with their host and in tune with the local environment. As long as we treat we get weaker bees who can only survive if we treat, and stronger parasites who can only survive if they breed fast enough to keep up with our treatments. No stable relationship can develop until we stop treating.

The other problem, of course, is that if we just stop now with the system of beekeeping we have, the genetically and environmentally weakened bees will usually die. Even if they are genetically capable of surviving in a clean (uncontaminated) environment, we have to get to an environment they can survive in or they will still die. So what is that environment?

Clean Wax

We need clean wax. Using foundation made from recycled, contaminated wax will not get that for us. The entire world wax supply is now contaminated with acaracides. Natural comb will provide clean wax.

Natural Cell Size

Next we beekeepers need to control the pests in a natural way. We will elaborate more on this as we go, but Dee and Ed Lusby arrived at the conclusion that the solution to this was to get back to natural cell size. Foundation (a source of contamination in the hive from pesticide buildup in the world beeswax supply) is designed to guide the bees to build the size cells we want. Since workers are from one size and drones from another and since beekeepers for more than a century have viewed drones as the enemy of production, beekeepers use foundation to control the size cells the bees make. At first this was based on natural sizes of cells. Early foundation ran from about 4.4mm to 5.05mm. But then someone (Francis Huber was the first) observed that bees build a variety of cell sizes and that large bees emerged from large cells and small bees emerged from small cells. So Baudoux decided that if you enlarged the cells more you could get larger bees. The assumption was that larger bees could haul more nectar and therefore would be more productive. So now, to-day, we have a standard cell size of foundation that is 5.4mm. When you consider that at 4.9mm the comb is about 20mm thick and at 5.4mm the comb is 23mm thick this makes a difference in the volume. According to Baudoux the volume of a 5.555mm cell is 301cubic mm. The volume of a 4.7mm cell is 192 cubic mm. Natural cell size runs from about 4.4mm to 5.1mm with 4.9mm or smaller being the common size in the core of the brood nest.

So what we have is unnaturally large cells making unnaturally large bees. We will elaborate more on why and how in the chapter *Natural Cell Size* in Volume II. The short version is that with natural cell size we get control of the Varroa population and can finally keep our bees alive without all the treatments.

Natural Food

Honey and real pollen are the proper food of bees. Sugar syrup has a much higher pH (6.0) than Honey (3.2 to 4.5) (Sugar is more alkali). Stating the same thing conversely, honey has a much lower pH than sugar syrup (Honey is more acidic). This affects the reproductive capability of virtually every brood disease in bees plus Nosema. The brood diseases all reproduce more at the pH of sugar (6.0) than at the pH of honey (~4.5). And this is not to mention that honey and real pollen are more nutritious than pollen substitute and sugar syrup. Artificial pollen substitute makes for short lived, unhealthy bees.

Learning

Newcomers in any field always seem to feel a bit overwhelmed, so before we get too far into this, let's talk about learning.

The most important thing you can learn in life is how to learn. I teach computer classes often and have always been a learner myself. I love to learn. I have discovered, though, that most people don't know how to learn. Here are some rules about learning that I don't think most people know.

Rule 1: If you're not making mistakes, you're not learning anything. I had a boss in construction who liked to say "If you're not making mistakes you're not doing anything." That may be true, but sometimes you are doing repetitious things and you can get to the point that you are not making mistakes, but if you are learning you will make mistakes! This is a fact. Making mistakes and learning are inseparable. If you're not making mistakes you're not pushing the limits of what you know, and if you're not pushing those limits, you're not learning.

My students in my computer classes often comment on how their children learn computers so quickly and easily and wish it was that easy for them. I tell them why it is easy for children. They are not afraid to make mistakes. Children are used to making mistakes. Adults are not. If you want to learn, get used to making mistakes. Learn from them.

I heard a story about a young man who was taking over as a bank president. The person who held the job before had been there for forty years and had made the company a lot of money. The young man asked him for advice before he left. The old man said that to make the bank money you make good decisions. The young man asked "how do you make good decisions?" The old man said, "you make bad decisions and learn from them." In the end, this is the really the *only* way to learn. Make mistakes and learn from them. I'm not saying you can't learn from other people's mistakes or from books, but in the end you have to make your own mistakes.

Rule 2: If you're not confused, you're not learning anything. If you are going to be a learner you will have to get used to being confused. Confusion is the feeling you get when you are trying to figure things out. Adults find this disconcerting, but there is no other way to learn. If you think back to the last card game you learned, you were told the rules, which you couldn't remember, but you started playing anyway. The first few hands were terrible, but then you started to understand the rules. But that was only the beginning. Then you played until you started to understand how to play strategically, but until you got good at it you were still confused. Gradually the whole picture of the rules and the strategies and how they fit together started to congeal in your mind and then it made sense. The only way from here to there, though, is that period of confusion.

The problem with learning and our world view is, we think things can be laid out linearly. You learn this fact, add this one and that one and then finally you know all the facts. But reality is not a set of linear facts; it is a set of relationships. It is those relationships and principles that understanding is made up of. It takes a lot of confusion to finally sort out all the relationships. There is no starting and ending point, because it is not a line, it is circles within circles. So you start somewhere and continue until you have the basic relationships.

Rule 3: Real learning is not facts, it is relationships. It's kind of like a jigsaw puzzle. You start somewhere, even though it doesn't look like anything yet. You sort things out by color and pattern and then you start fitting them together. Everything you learn in any subject is part of the whole puzzle and is related to everything else somehow.

The facts are just the pieces of the puzzle. You need them to figure out the relationships, but the pieces themselves don't make any sense until you have them connected. The connectedness of all things is one of the first things you need to learn in order to be able to learn.

A smart aleck news reporter once asked Albert Einstein how many feet were in a mile. Einstein said he had no idea. The news reporter then berated him, because he didn't know. Einstein said that's what he had books for, to look up things like that. He didn't want to clutter his mind with facts.

It is much more important to have a few facts and understand the relationships than lots of facts and no relationships. One little part of the puzzle put together is better than more pieces and none of them put together. Knowledge and understanding are not at all related. Don't go for knowledge; go for understanding, and knowledge takes care of itself.

Rule 4: It's not so important what you know as it is that you know how to find out. Tom Brown Jr. wrote a survival guide. I read survival guides all the time, but they usually frustrate me because they give recipes. Take this and that and do this with it and you have a shelter. The problem is, in real life you usually don't have one of the ingredients. Tom Brown, though, in his chapter on shelter, showed how he *learned* how to build a shelter. Telling you *how* to build a shelter and telling how to *learn* to build a shelter are as different as night and day. What you want to learn in life is not what the answers are, but how to find the answers. If you know that you can adjust to the materials and situations available.

The usual method is to look around and pay attention. Tom Brown learned to build a shelter by watching the squirrels, but he could have watched any animal that needed shelter and learned from them. Watching how other people and animals solve their problems and adapting those solutions is one way to learn.

Bee Basics

In order to do beekeeping, you need a basic understanding of their life cycle and their yearly "colony" cycle. You have two levels of organisms—the individual bee (which can't exist as an organism for very long) and the colony superorganism.

Lifecycle of a bee

Bees are one of three main castes: queen, worker or drone. The queen is the one bee that reproduces, but even that she can't do by herself. She is the one bee that goes out and mates, during one period of her life, that lasts a few days, and then she lays eggs for the rest of her life. The workers, depending on their age, feed brood, make comb, store honey, clean house, guard the entrance or gather honey, pollen, water or propolis. The drones spend their days flying out to drone congregation areas (DCAs) in the early afternoon and flying home just before dark. They spend their lives in hopes of finding a queen to mate with. So let's follow each cast from egg to death:

Queen

We will start with the queen since she is the most pivotal of any bee because there is generally only one of her. The reasons the bees raise a queen are: queenlessness (emergency), failing queen (supersedure), and swarming (colony reproduction).

Circle of Attendants

Queenlessness

The cells for each appear slightly different or at least occur under different conditions that can be observed. A queenless hive will have no queen that can be found, little open brood and no unhatched eggs. The queen cells resemble a peanut hanging on the side or bottom of a comb. If the queen died or was killed the bees will take young larvae and feed it extensive amounts of Royal Jelly and build a large hanging cell for the larvae.

Supersedure

In supersedure the bees are trying to replace a queen they perceive as failing. She is probably between 2 and 4 years old and not laying as many fertile eggs and not making as much Queen Mandibular Pheromone (QMP). These cells are usually on the face of the comb about $2/3$ of the way up the comb. There are, of course, exceptions. Jay Smith had a queen that was still laying well at 7 years named Alice, but three years seems to be the norm when the bees replace them.

Swarming

Swarm cells are built to facilitate the reproduction of the superorganism. It's how the colony starts new colonies. The swarm cells are usually on the bottom of the frames making up the brood nest. They are usually easy to find by tipping up the brood chamber and examining the bottom of the frames.

The larvae that make a good queen are worker eggs that just hatched, which happens on day $3^1/_2$ from the day the egg was laid. On day 8 (for large cell) or day 7 (for natural sized cells) the cell will be capped. On day 16 (for large cell) or day 15 (for natural sized cells) the queen will usually emerge. On day 22, weather permitting, she may fly. On day 25, weather permitting, she may mate over the next several days. By day 28 we may see eggs from a new fertile queen. From that time on, she will lay eggs (weather and stores permitting) until she fails or swarms to a new location and starts laying there. The queen will live two or three years in the wild, but almost always fails by the third year and is replaced by the workers. In a swarm the old queen leaves with the first (primary) swarm. Virgin queens leave with the subsequent swarms, which are called afterswarms

Worker

Worker Bee Gathering Propolis

A worker egg starts out the same as a queen egg. It is a fertilized egg. Both are fed royal jelly at first, but the worker gets less and less as it matures. Both hatch on day $3^1/_2$ but the worker develops more slowly. From day $3 \ ^1/_2$ until it is capped it is called "open brood". It is not capped until the 9th day (for large cells) or the 8th day (for natural sized cells). From the day it is capped until it emerges it is called "capped brood". It emerges on the 21st day (for large cells) or the 18th or 19th day (for natural sized cells). From when the bees start chewing through the caps until they emerge they are called "emerging brood". After emergence a worker starts its life as a nurse bee, feeding the young larvae (open brood). For those who say that a worker is an incomplete female while a queen is a fully functioning female, consider that only a worker can produce "milk" for the young. Only a worker can feed and care for the young. The queen does not have the right glands to produce food for young, nor the skills to care for them. Neither the worker nor the queen is a "complete mother"; it takes them both to rear the young. Workers and queens are anatomically different in many ways. Only a worker has the hypopharyngeal gland to feed the young. Only a worker has baskets for carrying pollen and propolis. Only a queen can lay fertile eggs. Only a queen can make sufficient pheromones to keep the hive working correctly.

For the first 2 days the newly emerged worker will clean cells and generate heat for the brood nest. The next 3 to 5 days it will feed older larvae. The next 6 to 10 days it will feed young larvae and queens (if there are any). During this period from 1 to 10 days old it is a Nurse Bee. From day 11 to 18 the worker will make honey, not gather but ripen nectar and take it from field bees bringing it back, and build comb. From days 19 to 21 the workers will be ventilation units and guard bees and janitors cleaning up the hive and taking out the trash. From day 11 to 21 they are House Bees. Day 22 to the end of their life they are foragers. Except during winter, workers usually live about six weeks or less, working themselves to death until their wings are too shredded to fly. If the queen fails a worker may develop ovaries and start to lay. Usually these are drone eggs and usually there are several to a cell and they are in worker cells.

Drone

Drones are from unfertilized eggs. For those of you who studied any genetics, they are haploid, meaning they only have a single set of genes, where a worker and a queen are diploid, meaning they have pairs of genes (twice as many). Drones are larger than workers but proportionately wider, shorter than a queen, have a blunt back end, have huge eyes and no stinger. The egg hatches on day $3^1/_2$. The cell is capped on day 10 (for large cells) or as early as day 9 (for natural sized cells) and emerges on day 24 (for large cells) or between day 21 and 24 (for natural sized cells). The colony will raise drones whenever resources are plentiful so that there will be drones to mate with a queen if they are needed. It is unclear what other purposes they serve, but since a typical hive raises 10,000 or more of them in the course of year and only 1 or 2 ever get to mate, they may serve other purposes. If there is a shortage of resources the drones are driven out of the hive and die from cold or starvation. The first few days of their lives they beg food from the nurse bees. The next few days they eat right from the open cells in the brood nest (which is where they usually hang out). After a week or so they start flying and finding their way around. After about two weeks they are regularly flying to DCAs (Drone Congregation Areas) in the early afternoon and stay until evening. These are areas where drones congregate and where the queens go to mate. If a drone is "fortunate" enough to mate, his reward is to have the queen clamp down on his member and rip it out by the roots. He will die from the damage. The queen stores up the sperm in a special receptacle (spermatheca) and distributes it as she lays the eggs. When the queen runs out of stored sperm, she does not mate again, she fails and is replaced. I think drones have an undeserved reputation for being useless. In fact they are essential. Not only do they have a reputation for being useless but for being lazy. They are not lazy. They fly until they are exhausted every day that the weather permits, trying to ensure the continuation of the species.

Yearly cycle of the colony

By definition this is a cycle so we'll start when the year really begins, in the winter. I can speak to what happens in Nebraska. For your location I would consult local beekeepers.

Winter

The colony tries to go into winter with sufficient stores, not only to survive the winter, but to build up enough by spring for the colony to reproduce. To do this the colony needs a good supply of honey and pollen. The bee colony appears to be dormant all winter. They usually don't fly unless the temperatures get up around 50° F (10° C). But actually the bees maintain heat in the cluster all winter and all winter the colony will rear little batches of brood to replenish the supply of young bees. These batches take a lot of energy and the cluster has to stay much warmer during them. The colony takes breaks between batches. As soon as there is any supply of fresh pollen coming in the colony will begin buildup in earnest. Usually the early pollen is the Maples and the Pussy Willows. In my location this is late February or early March. Of course if the weather isn't warm enough to fly, the bees won't have any way to get it. Beekeepers often put pollen patties on at this time so the weather won't be a deciding factor in the buildup.

Spring

By spring the colony is now building up well. They should have raised at least one turnover of brood by now. They will really take off with the first bloom. This is usually dandelions or the early fruit trees. Here in Nebraska, that's the wild plums and chokecherries which will bloom about mid April. Between now and mid May the colony will be intent on swarm preparations. They will try to finish building up and then start back filling the brood nest with nectar so the queen can't lay. This sets off a chain reaction that leads to swarming. The more the queen doesn't lay the more she loses weight so she can fly. The less brood there is to care for, the more unemployed nurse bees there are (the ones who will swarm). Once critical mass of unemployed nurse bees is reached, they will build swarm cells, the queen will lay in them and the colony will swarm just before they are capped. All of this is assuming, of course, that there are abundant resources and that the beekeeper doesn't intervene. If they decide not to swarm then they go full throttle into nectar collection. If they decide *to* swarm then the old queen leaves with a large amount of the young bees and try to start a new home somewhere. Meanwhile the new queen emerges in a couple of weeks and starts laying in another couple of weeks and the remaining field bees haul in the crop to build up for the next winter.

Summer

Our flow, here in Nebraska, is really mostly in the summer. This is usually followed by a summer lull. It seems to be driven, here in my location anyway, by a drop in rainfall. Sometimes if the rain is timed right there isn't really a lull at all, but usually there is. Our flow starts about mid June and ends when things dry up enough. Sometimes there's an actual dearth where there is no nectar at all and the queens stop laying. I'd say most of my nectar is soybeans, alfalfa, clover, and just plain weeds. This varies greatly by climate.

Fall

We usually get a fall flow in Nebraska. It's mostly smartweed, goldenrod, aster and chicory with some sunflower and partridge pea and other weeds. Some years it's enough to make a crop. Some years it's not enough to get them through the winter and I have to feed them. Around mid October, usually, the queens stop laying and the bees start settling in for the winter.

Products of the hive

The bees produce a variety of things. Most of these are gathered from the bees by people.

Bees

Many producers raise bees and sell them. Package bees are available from the Southern United States usually in April.

Larvae

Many people over the world eat bee larvae. It is not that popular here in the US. To raise larvae (which the bees have to do to get bees) the bees need nectar and pollen. Feeding syrup or honey and pollen or pollen substitute is a way to stimulate the bees in the spring to raise more brood and therefore more bees.

Propolis

The bees make this from tree sap that is processed by enzymes the bees make and mix with it and sometimes they mix in beeswax. The substance most often gathered is from the buds of relatives of the poplar family, such as poplar, aspen, cottonwood, tulip poplar and others. It is used in the hive to coat every-thing. It is an antimicrobial substance and is used both for sterilizing the hive and for structural help. Every-thing in a hive is glued together with this. Openings that the bees think are too big are closed with this. Humans use it as a food supplement and as a topical anti microbial for cuts and for cold sores etc. It kills both bacteria and viruses. Propolis traps are available. A simple one is a screen over the top of the hive and you roll it up and put it in the freezer and then unroll it while it's frozen to break all the propolis off.

Bee Gathering Propolis Gathering Propolis from old equipment

Wax

Anytime a worker bee has a stomach full of honey and nowhere to store it, it will begin to secrete wax on its abdomen. Most of the wax is then used to build comb. Some falls on the floor of the hive and is wasted. For humans, beeswax is edible, although it has no nutritional value. It is used in foundation, candles, furniture polish and cosmetics. The bees need it to store their honey in and raise their brood in. To get it from the bees, either crush comb and drain the honey, or use cappings from extracting and melt and filter them.

Pollen

Pollen has a lot of nutritive value. It is high in protein and amino acids. It is popular as a food supplement and is believed by many to help with their allergies, especially if it is pollen collected locally. The bees need it to feed the young. Pollen traps are available commercially or you can find plans to build your own. The principle of a pollen trap is to force the bees through a small hole (the same as #5 hardware cloth) and in the process they lose their pollen which falls into a container through a screen large enough for pollen but too small for the bees (#7 hardware cloth). Some pollen traps must be bypassed about half the time so the hive doesn't lose its brood from lack of pollen to feed the brood. A week on and a week off seems to work. Other problems with pollen traps is drones not getting access in and out and if a new queen is raised, she has difficulty getting out and can't get back in. If you are allergic and trying to treat allergies with pollen take it in very small doses until you build a tolerance or until you have a reaction you don't want. If you have a reaction either take less or none at all depending on the severity.

(Photo by Theresa Cassiday)

Pollination

A "product" of having bees is that they pollinate flowers. Pollination is often a service that is sold. $50 to $150 (depending on the supply of bees) for $1^1/_2$ deep boxes is a typical charge for pollination. Pollination charges are usually based on having to move the hives in and out in a specific time frame so that the trees (or other plants) can be sprayed etc. It is less likely there will be charges for pollination if the bees can be left there year round and pesticides are not used. In this case it is usually a mutually beneficial situation for the beekeeper and the farmer and there usually is no charge or rent either way, although it's common for the beekeeper to give the farmer a gallon of honey from time to time.

Honey

This is what is usually considered the product of the hive. Honey, in whatever form, is the major product of the hive. The bees store it for food for the winter and we beekeepers take it for "rent" on the hive. It is made from nectar, which is mostly watered down sucrose, which is converted to fructose by enzymes from the bees and dehydrated to make it thick.

Honey is usually sold as Extracted (liquid honey in a jar), Chunk comb (a chunk of comb honey in liquid in a jar), Comb honey (honey still in the comb. Comb honey is done in Ross Rounds, section boxes, Hogg Half combs, cut comb, and more recently Bee-O-Pac. It is also sold as creamed honey (where it is crystallized with small crystals).

Since the subject always comes up, all honey (except maybe Tupelo) eventually crystallizes. Some does this sooner and some later. Some will crystallize within a month; some will take a year or so. It is still edible and can be liquefied by heating it to about 100 degrees or so. Crystallized honey can be eaten as is also, or crushed to make creamed honey or feed to the bees for winter stores. It crystallized most quickly and therefore most smoothly, at 57º F. The closer it is stored to that temperature, the more quickly it will crystallize.

Royal Jelly

The food fed to the developing queen larvae is often collected in countries where labor is cheap and sold as a food supplement.

Four Simple Steps to Healthy Bees

I touched briefly on this in the chapter *Why this Book* but we will go into more depth here.

For the moment, let's look at just these four issues: comb; genetics; natural food; and no treatments. Let's gloss over the arguments and focus only on what we know to be facts and what we can do about them.

Comb

I find all the arguments over cell size and whether it does or does not help your Varroa issues and all the rest a bit tiresome. Varroa is no longer an issue in my beeyards and yet I find that the obsession of every bee meeting I go to seems to be Varroa about half of what I end up talking about is Varroa. I went to natural cell and small cell at a time when no one believed it was possible to keep bees alive without treatments. After doing no treatments with repeatedly disastrous results before, I came to the same conclusion. But after going to small and natural cell size I was pleased to be back to keeping bees instead of managing mites. This anecdotal evidence is not enough for some, even as the same from others was not enough for me until I tried it, but unlike me they don't seem to be willing to try it. But let's consider your choices:

You can assume that cell size is irrelevant to everything, if you like. This seems like a doubtful assumption since we know for a fact it has everything to do with the size of bees. If scaling up the entire body of a bee to 150% of what it was naturally is not a significant change, then I don't know what you would consider significant. We've known this is a fact since Huber's observations and in addition we have reams of research by Baudoux, Pinchot, Gontarski and others as well as recent research by McMullan and Brown (The influence of small-cell brood combs on the morphometry of honeybees (Apis mellifera)—John B. McMullan and Mark J.F. Brown).

Choices

Natural Cell Size

You can assume whatever you like about what size *is* natural. But in the end the only way to get natural cell size, and let the bees end the debate, is to stop giving the bees foundation and let them build what they want. Since that is what bees do if you let them and since that is actually less work for you than using foundation and less expense and since that's the only way to get uncontaminated combs (do an online search for the video of Maryann Frasier on contamination by acaracides in new foundation) it seems like a win-win-win to me. Even allowing the assumption that cell size is irrelevant, no one is saying that natural cell size is bad for the bees and no one I know of thinks that clean wax is bad for the bees and most are very convinced at this point that clean wax is essential for truly healthy bees.

Why not let them build what they want?

Why wouldn't you let them build what they want? It seems there is a lot of fear that the bees will only build drones. I have heard this from many beekeepers. Obviously this is not true. If it were there would never have been any feral bees. If you want to know how much drone comb they will build and how many drones they will raise and how much influence you can have on it, read Clarence Collison's research on the subject (Levin, C.G. and C.H. Collison. 1991. The production and distribution of drone comb and brood in honey bee (Apis mellifera L.) colonies as affected by freedom in comb construction. BeeScience 1: 203-211.). The point is that in the end the amount of drones is controlled by the bees and leaving them that control in the first place will simplify life for them and you. The thing to do when the bees draw a frame full of drone comb in the middle of the brood nest, is set it to the outside edge of the box and give them another empty frame. Otherwise, if you take it out, their need for drone comb and drones unfulfilled, they will draw yet another frame of drone and contribute to the myth that if you let them, they will draw nothing but drone comb.

Combs in frames?

Another fear seems to be that the bees will not draw the combs in the frames. They will mess up foundationless about the same rate as they mess up any other system of foundation. They will mess up plastic foundation a lot more than foundationless frames. But if they do, you just cut it loose and tie it into the frame, if it's brood, or harvest it, if it's honey.

Draw comb without foundation?

I've even heard old timers tell new beekeepers that without foundation the bees won't draw comb at all. This is so patently absurd that I don't see any need to respond to it.

Wire?

The last seems to be the myth that wire is necessary in order to extract. The wire was added to foundation to keep the foundation from sagging before it was drawn (see any older ABC XYZ of Bee Culture). It was not added to allow extraction. Extraction is done on unwired foundationless frames by many people, including me. But if wire is your hang-up, add some wire to the frames, level the hive and sleep well. I prefer to just use mediums and be able to lift the boxes and have had no need at all for the wires.

How do you do foundationless?

- With standard wedge frame, just break out the wedge and nail it sideways.
- With grooved top bars, put popsicle sticks in the groove or a half of a paint stick or a piece of a "one by" ripped
- With drawn wax, just cut the center of the comb out leaving a row of cells around the edges

- With an old frame with no comb, just put it between two drawn brood combs
- With a plastic foundation/frame, just cut the center of the foundation out leaving a row of cells around the edge
- When making your own, cut a bevel on the top bar so it slopes down to a point. You can also make them $1^{1}/_{4}$" wide.

Less work

So how much work is foundationless? We talked about *how* to do it, but how much work is it? If you buy standard wedge frames and turn the wedge 90 degrees and glue and nail it back on you have a foundationless frame. That is pretty simple. You were going to break it out and nail it in anyway weren't you? The other methods above were less work than wiring wax foundation. The only slightly tricky thing would be plastic frames with built in foundation. Then you'd need to cut the center of the foundation out. That could be done with a number of tools, but I suppose a really hot knife would cut it out pretty quickly. A jig and a router would probably do ok as well and it would be simple to leave the corners and edges in for strength and for a guide. So how does this compare with putting in wire, crimping, foundation, embedding etc.? Or using plastic? You save as much as $1 a sheet if you wanted to get small cell or close to that if you wanted to get plastic.

Downside?

So, for less work and less money you can end up with clean wax, natural cell size and a natural brood nest as far as distribution of cell sizes and drones. What's the down side? If you don't wire the deeps you might end up with more collapsed comb if you have a migratory operation, because of bumpy roads combined with hot days and deep frames, but you could wire them and that would probably not be so much of a problem. You would also need to keep the boxes more level, which in a fixed operation isn't so hard; you just level the stands up, which you should have done anyway. But in a migratory operation it would take more work to level them than to just set the pallets down and not worry about them being level.

Timeline

Worst case timeline is you retool at whatever pace you would have done by the other method anyway. You buy foundation and put it in all the time, right? Some rotate their comb out every five years or less. Some just replace comb as they need comb but either way if you stop using the large cell foundation and stop treating you'll eventually have natural clean comb by the only possible method to get clean comb unless someone finds a source of clean wax and makes their own foundation.

If you have a lot of large cell foundation around, you can sell it to someone local who was going to buy some anyway for the catalog price and save them the shipping. Or, if you're impatient, sell it cheap, if you're willing to take a small loss for healthier bees. You can make up the difference on all those strips that weren't working anyway that you won't have to buy.

Worst case scenario

So let's look at worst case scenario. Let's assume that cell size isn't an issue one way or the other. It's unreasonable to assume that bees will be any *less* healthy on natural sized comb, so at worst they will be on a cell size no better. At worst the cost is less than rotating out your contaminated combs for contaminated wax foundation. There is hardly a down side to that. The *work* is less than wiring wax foundation. The *cost* is less than wiring wax foundation. The wax will be uncontaminated (at least unless or until *you* contaminate it) and we *know* that wax contamination is contributing to lack of longevity and fertility in queens and drones. So we know the bees will be healthier and the queens will do better.

Best case scenario

This is the worst case scenario on all of the speculation on cell size and natural comb. The best case scenario is that it will solve your Varroa problems.

No Treatments

I don't know what all the rest of you have experienced, but with no treatments (on large cell size) I lost all my bees whenever I wouldn't treat for a couple of years. But finally I lost them even after treating with Apistan. It was obvious that the mites had built resistance. I've heard of big outfits losing their entire operation *while* treating with Apistan or CheckMite. So we have reached the point where whether you treat or not, they all die anyway quite often. I think the problem here comes down to us not wanting to "do nothing". We want to attack the problem and so we do whatever the experts tell us because we are desperate. But what they are telling us is failing anyway. Once I lost them all *after* I treated them, I could no longer see any reason to treat them. Treating only perpetuates the problem. It breeds bees that can't survive whatever you are treating for, contaminates the comb and upsets the whole balance of the hive.

Ecology of the hive

There is no way to maintain the complex ecology of a natural beehive while dumping in poisons and antibiotics. The beehive is a web of micro and macro life. There are more than 30 kinds of benign or beneficial mites, as many or more kinds of insects, 8,000 or more benign or beneficial microorganisms that have been identified so far, some of which we know the bees cannot live without and some of which we suspect keep other pathogens in balance. Every treatment we dump in a hive, from essential oils (which interfere with the bees smell, which is how everything in the dark of the hive is communicated, and kill microorganisms, beneficial and otherwise); to organic acids (which kill microorganisms as well as many insects and benign mites) to acaracides (which are always just chemicals that kill arthropods which include insects and mites but kill mites at a slightly higher rate); to antibiotics (which kill the microflora most of which is either beneficial or benign but useful in maintaining the balance and crowding out pathogens); even to sugar syrup (which has a pH that is detrimental to the success of many of the beneficial organisms and advantageous to many of the pathogens: EHB, AFB, Chalkbrood, Nosema etc. unlike the pH of honey that is much lower and detrimental to the pathogens and hospitable to many known beneficial organisms). I think we've reached the point that it's silly to act like we've been doing any good when the bees are collapsing in spite of, if not because of all of this.

Downside of not treating

So what is the downside of not treating? Worst case is they die. They seem to be doing that regularly enough already aren't they? I don't see that I'm contributing to that by giving them the chance to reestablish a naturally sustainable system. I'm just not destroying that system arbitrarily to get rid of one thing with no regard to the balance of the system. Of the people I know who are not treating for anything, even on large cell; their losses are *less* than those who *are* treating. On small cell or natural cell they are even less. But even if you don't buy the cell size debate, not treating is working as well as treating is. I go to bee meetings all over the country and hear people who, like me, lost their bees when they were treating religiously and then decided to just stop. Their new bees are now doing better than when they were treating them. I feel bad when I see a dead hive, but I also say "good riddance" to the genetics that couldn't make it.

If you think you'll have too many losses (my guess is you already *do* have too many losses) and you can't take those losses, what would it take to make splits and overwinter enough nucs to make up those losses every spring with your own locally adapted stock? A bunch of walk away splits made in the middle of July, after cashing in on the main flow, will usually winter, at least around here, and not put a dent in your honey crop. You can also split the mediocre hives earlier since they weren't doing much anyway, requeen with cells from your best stock and not really affect your honey crop You can also do cut down splits on the strong hives right before the main flow and get good splits, well fed queens, more honey *and* more hives.

Upside of not treating

What is the upside of not treating? You don't have to *buy* the treatments. You don't have to *drive* to the yard and put the treatments in and *drive* to the yard to take them out. You don't have to contaminate your wax. You don't upset the natural balance by killing off micro and macro organisms that you weren't targeting but who are killed by the treatments anyway. That would seem like upside enough, but you also give the ecosystem of the bee hive a chance to find some natural balance again.

But the most obvious up side is that until you quit treating you can't breed for survival against whatever your issues are. As long as you treat you prop up weak genetics and you can't tell what weaknesses they have. As long as you treat you keep breeding weak bees and super mites. The sooner you stop, the sooner you start breeding mites adapted to their host and bees who can survive with them.

Breeding locally adapted queens

Breeding locally adapted queens from the best survivors is another thing that I don't see a downside to. If you breed from your untreated survivors you'll get bees that are surviving where you are against what they face there. They will mate with the local ferals who are also surviving. The propaganda that you can't raise queens that are as good as or better than commercially available queens is just that—propaganda. The same is true with the need to requeen early in the spring. Early queens are often not well mated and often not well fed. Assuming you don't treat, you don't requeen regularly and you use your most successful survivors, your queens are more likely to be better because of the following:

- They are locally adapted.
- They are bred from survivors.
- You can raise them at optimum times to have plenty of nutrition and plenty of drones.
- They are probably never caged and go from laying in the mating nuc to the hive they are put in with no break. This develops better ovarioles and that makes

better pheromones. This results in them be more long lived, laying better patterns, swarming less and being accepted better.

- You save a lot of work. If you keep queens longer and mate from those that succeed at superseding at appropriate times you have bees that can requeen themselves. This will save you a lot of labor in finding queens and introducing queens as the bees will take care of this.
- Even on the hives you requeen, you can save labor by requeening with cells and not bothering to find the old queen. The new queen will typically be accepted and you didn't have to spend the day looking for the old one.
- You save a lot of money. Open mated production queens go for from $15 to $40 and breeders go for much more.
- You can easily keep spares in nucs and have queens whenever you need them.

What about AHB?

Those in AHB areas seem concerned about this approach. I'm not in such an area, but it seems to me that ancestry isn't my concern. Temperament is. Productivity is. Survival is. If you only keep the gentle ones and requeen the hot ones I think it will work fine. Those I know doing this in AHB areas have come to that conclusion. Another thing to consider is that F1 crosses are often hot. So if you keep bringing in outside stock you may be contributing to them being angry. You may be better off selecting for gentleness and requeening any hives that are hot with local stock that is gentle.

Natural Food

It's quite simply less work to use natural food. If I don't feed pollen substitute in the spring then I don't have to make patties etc. If I don't feed syrup, I don't have to buy sugar, I don't have to make syrup, I don't have to drive to the yards and I don't have to feed it. If I leave them honey to winter on, there is less honey for me to pull, haul home, extract, haul back empty to get cleaned up and then pull off to store, make syrup, drive to the yards to feed it etc. This is less work all the way around. Even if you don't believe that honey is more nutritious to bees (although I have to wonder why you want to produce honey if you think there is no difference between honey and sugar). It is definitely less work to leave it. Even if you believe that the difference in pH is irrelevant (which I seriously doubt), it's less work than making syrup and feeding syrup. Even if you are obsessed with the difference in price ($0.40 per pound for sugar vs. some variable price from say $0.90 to $2.00 pound for honey) by the time you extract the honey, buy the sugar, make the syrup, haul it to the yards, feed it, go back and pull the feeders etc. do you honestly think you came out that far ahead? It's not just a $0.60 a pound difference by the time you factor all of that in, unless your labor is of no value. So let's assume that the difference in the health of the bees is only marginal between honey and sugar and ignore that Nosema multiplies better at the pH of sugar than honey and so does Chalkbrood and EFB and AFB. We'll ignore all of that and just assume it's marginal. If there is ANY difference it could tip the scale from a colony surviving and one dying and packages are up around $80 delivered here.

Looking more into pH

Sugar syrup has a much higher pH (6.0) than Honey (3.2 to 4.5) (Sugar is more alkali). Conversely, honey has a much lower pH than sugar syrup (Honey is more acidic). This affects the reproductive capability of virtually every brood disease in bees plus Nosema. They all reproduce better at pH 6.0 than at 4.5.

Chalkbrood as example

"Lower pH values (equivalent to those found in honey, pollen, and brood food) drastically reduced enlargement and germ-tube production. Ascosphaera apis appears to be a pathogen highly specialized for life in honeybee larvae." —Author. Dept. Biological Sci., Plymouth Polytechnic, Drake Circus, Plymouth PL4 8AA, Devon, UK. Library code: Bb. Language: En. Apicultural Abstracts from IBRA: 4101024

Similar information is available concerning other bee diseases. Try an internet search for pH and AFB or

EFB or Nosema and you'll find similar results on their reproductive capability related to the pH.

Differences in pH affect other beneficial and benign organisms in the hive. The other more than 8,000 microorganisms in the hive are also affected by changes in pH Using sugar syrup also disrupts the ecological balance of the hive by disrupting the pH of the food in the hive and the food in the bees' gut.

Pollen

If you don't use pollen substitute you can still leave pollen in the hives and if you really want you can set aside a hive or two or more (depending on the size of your operation) and trap a few pounds of pollen to put in an open feeder in the spring. Just freeze it in the meantime. I put it on a screened bottom board on top of a solid bottom board with an empty box on top with a lid. The screen keeps the bottom dry and the hive keeps it from getting rained on.

Pollen trapping

The cost of trapping is mostly the trap. If you do it in a yard close to or on the way home it's easy enough to empty the traps every night. And now you don't have to buy pollen patties and you have superior nutrition.

If you doubt the difference, look for research on bee nutrition that compares substitutes to pollen. Bees raised on substitutes are short lived and weak.

Synopsis

So what do you have to lose? You can get better genetics for your bees by breeding your own; cleaner comb by using foundationless and no treatments; longer lived bees from clean wax and feeding real pollen; and less work by leaving honey that you won't have to harvest and feed syrup back; and the worst case is that to get all this you'll work less and the best case is that it will all have a markedly positive effect on the health of your bees. Worst case, if you implement this a little at a time, you lose some bees, which you're already doing. Best case you lose less.

Different Profit Formula

Let's try a different profit formula. How much time, gas, work, and money do you spend on syrup, feeding, putting in patties, putting in treatments, taking out treatments, harvesting that last little bit of honey that you then have to make up with syrup, putting in foundation etc.? How much money and time would you save if you stopped doing all of that? How many more hives could you handle and how much more honey would they make?

Choices

Too Many Choices?

I realize many people simply want someone to tell them to do "a" "b" and "c" and it will work for them. I also realize that in the context of a beginner this may be the best directions you can give, but on the other hand I have never appreciated that kind of "one size fits all" advice and have always preferred to know what my options were. Perhaps I overwhelm the newcomers with too many options, but on the other hand I don't feel I can say there is only one right answer when there really isn't. Perhaps I should leave out the things I've left behind, but I have an assortment of things I'm still using and it's difficult to say that one is better or worse than another when there are appealing things about them all.

Beekeeping Philosophy

Some of those options are related to your philosophy and your energy. In these examples I will assume you want to get to natural cell size or small cell size and no treatments. So, for instance, if you just can't handle the idea of plastic, then there is no point in considering Honey Super Cell or Mann Lake PF120s or PF100s or PermaComb or PermaPlus as options. You may as well just limit yourself to wax 4.9mm foundation or foundationless. But if plastic does not run contrary to your view of life, the PF120s will save a lot of labor over building foundationless frames, and a lot of cost over Honey Super Cell. So knowing you have that option might be helpful to you in making your choice.

Time and Energy

More on the energy and time front, if you have the energy and time, I like to cut my frames down to $1^1/_4$" instead of the standard $1^3/_8$" but it takes time and energy and tools. So I have a lot of Mann Lake PF120s that are standard width and probably will never get time to cut them down.

Feeding Bees

This also carries over to feeders and other things. For instance, having hive top feeders that hold five gallons is nice for feeding an outyard in early fall, but is also expensive. Feeding hives in my back yard can work fine with bottom board feeders (that cost me nothing) and more frequent trips. Having these options doesn't mean one is better than the other, but one may fit your situation better than the other. Buying feeders for 200 hives is not practical for me so I feed my outyards when necessary, with dry sugar in empty boxes. They tend to eat it but not store it. This saves me buying feeders, making syrup and save the bees having combs full of sugar syrup and me having to keep track of that so I don't harvest sugar syrup. Is that the best solution? It seems to work well for me, but may or may not work well for you.

Take your time

My point is that options, in my opinion, are good, but they also sometimes create a lot of overwhelming decisions for a new beekeeper who has no frame of reference for those decisions. One good step is grow slowly in your beekeeping and don't invest too heavily in anything that is special equipment until you've had time to test it thoroughly. Most beekeepers have wasted a lot of money on equipment they eventually didn't use. Of course part of this may be to see what you can get by without, instead of trying out everything on the market. For example, feeding with an empty box and dry sugar is much cheaper and less investment than buying hive top feeders.

Important Decisions

One of the most important things to do is sort out the hard to change decisions from the less important, easy to change decisions.

If you pay attention to the rest of this you'll see that hardly anything I *would* buy is in a beginner bee-keeper starter kit.

There are many things in beekeeping you can easily change as you go along. There is no point stressing out over these things. There are other things in beekeeping that are an investment and are difficult to change later.

Easy Things to Change in Beekeeping:

You can always go to a top entrance. You only have to block the bottom one (with a $^3/_4$" by $^3/_4$" by $14^3/_4$" entrance block on a ten frame standard bottom board) and propping up the top. It's not like everything you have is outdated if you decide that you want a top entrance.

You can always choose to put in or leave out a queen excluder. Odds are, sooner or later, you'll need one for something. They are handy for the bottom of an uncapping tank or as an includer when hiving a swarm etc. It's not that big of an investment to have one or two (or not). Nor is it that big of a problem to buy one later if you don't have one.

You can change the race of bees *very* easily. You'll probably requeen once in a while even if you *aren't* trying to change races, and all you have to do is buy a queen of whatever race you want and requeen. So it's not that critical what breed you pick. I doubt you'll be disappointed with an Italian or a Carni or a Caucasian. And if you decide you want something else, it's not hard to change.

Difficult Things to Change in Beekeeping:

The bigger issues are things that are an investment you have to live with or you have to go to a lot of trouble to modify or undo.

If you think you want small cell (or natural sized cell) you're one step ahead to use it from the start. Otherwise you'll have to either gradually phase out all of the large cell comb or do a shakedown and do it all at once. If you invested money in plastic foundation, this is disappointing (I have hundreds of sheets in my basement of large cell foundation I'll never use). But at least you won't have to cut down all your equipment.

If you buy a "typical" starter kit you'll get ten frame deeps for brood and shallows for honey. The ten frame deeps full of honey weigh 90 pounds. Some will argue that when they have brood in them they weigh less than that. That's true. But sooner or later you'll have one full of honey and you may not be able to lift it. If you go with all mediums you'll have to be able to lift 60 pound supers full of honey. If you go with eight frame mediums you'll only have to lift 48 pound boxes. I started off with the deep/shallow arrangement and had to cut down every box and frame to mediums. Then I cut all the ten frame boxes down to eight frames. It sure would have been easier to just buy eight frame mediums from the start. Interchangeability is also a wonderful thing.

Screened bottom boards are easy to just buy. It's harder to convert the old ones.

If you buy a lot of *anything*, you may decide you hate it later. Make changes slowly. Test things before you invest a lot in them. Just because one person likes it, doesn't mean you will like it.

Choices I recommend

So, if you want to minimize your choices and maximize your success I'll distill things down to what I would recommend with only a few choices:

Frame depth

I'm going to recommend you use all the same size frames for everything, and since medium frames seem like the best compromise for everything, I'm going to recommend mediums for everything, mainly because of lighter boxes. That includes comb honey, extracted honey, brood etc. These are sometimes called Illinois supers. Or $^3/_4$ supers. They are $6^5/_8$″ deep with $6^1/_4$″ frames.

Reasons for all the same size: You can bait up supers with brood, or other frames from the brood chamber. You can pull honey from the supers for starting nucs etc. You can run an unlimited brood nest and if the queen lays in the supers, you just pull those frames of brood and swap them for some honey from the brood chamber. Different sizes are really a deterrent to good management of the hive.

Reasons for mediums instead of deeps: A 10 frame deep full of honey can weigh up to 90 pounds. A medium full of honey can weigh up to 60 pounds. 'Nuff said.

Various frames from extra shallow to Dadant deep

Various depths of boxes from deep to extra shallow

Number of Frames

Now that we have a frame size you need to pick a hive size. Standard is 10 frames. There is much to be said for being standard. On the other hand, there is much to be said for lighter (48 pounds vs. 60 pounds). The 8-frame equipment from Brushy Mt. or Miller Bee Supply or Walter T. Kelley or others, is very nice for making less work. You need to choose whether you want lighter boxes or standard sized ones. I converted to 8 frame. One of the other advantages of 8 frame equipment is that it is such a more versatile size. It is the same volume as a 5 frame nuc and can be used for a nuc. With a follower board it could even be used for a 2 frame mating nuc and then expanded, if need be, to eight frames eventually.

Various widths of boxes from two frames to ten

Style of Frames and Cell Size of Foundation

Frames, foundation, cell size etc. You need to decide if you want plastic foundation, plastic frames, fully drawn plastic comb, etc. and what size you want the foundation. I would recommend just buying small cell or PermaComb or Honey Super Cell. If you want to use wax, buy small cell wax from Dadant or one of the other suppliers. The small cell plastic is no longer on the market from Dadant. But Mann Lake's PF120's are 4.95mm cell size and are one piece frame and foundation. If you want to not have to build frames, not have to wait for the bees to draw it and never have to worry about wax moths or Small Hive Beetles then buy PermaComb or Honey Super Cell. I personally heat the PermaComb to 200º F and dip it in 212º F beeswax and shake off all the excess wax. This results in 4.9mm cells and seems to handle all my mite problems. For now don't worry about regression or all that complex sounding stuff, but just stick with natural sized or small cell (aka 4.9mm) foundation. Or use foundationless (see that chapter for more information).

Eight Frame Mediums

Left to right, eight frame, ten frame, eight frame

To minimize injuries from lifting and make life simple, buy all eight frame medium boxes. Pick a manufacturer who is reasonable in price and shipping to your location.

Plastic Small Cell Frames

If you don't mind plastic, buy all Mann Lake PF120 frame/foundation so you don't have to learn to (and find time to) build frames, wire foundation etc. These have been the most successful at getting small cell comb right off the bat in my experience.

If you don't like the idea of plastic

Then use foundationless. Certainly foundationless is the most appealing to me as you can't get any more natural than that. I would buy the wedge top bar frames and rotate the wedge 90 degrees so it makes a comb guide.

Jay Smith style Bottom Feeder

Bottom Board Feeders

I would buy solid bottom boards and convert them into bottom board feeders. There is no reason to spend a lot of money on feeders if your management plan is to leave them honey instead of feeding and only feed in emergencies.

I would make those feeders the style with no entrance and a plug for a drain and build simple top covers with top entrances to eliminate skunk, mice, grass, snow and condensation issues.

Essential Equipment

Here are some essentials for the beekeeper:

Large Smoker
I would buy a good smoker. A large one. Large ones are easier to light and keep lit. Smaller ones are harder to light and keep lit. I would light the smoker anytime you are going to do more than just pop the top and I would light it most of the time even then if there is a dearth or any other reason to suspect they might be defensive. Don't oversmoke them. Make sure it's lit well and put a puff in the entrance and after you open up a puff across the top bars. Put the smoker down and leave it unless they start to get excited.

Veil, jacket, or suit
I would prefer, if I only have one protective suit, to have a jacket with a zip on veil. It's what I use the most, but it is nice to have a full coverall with a zip on veil. That way I can be pretty fearless of the bees. If you make them mad enough, long enough, they will still get in, but that would require quite a bit of time. If you have the money to spare, I'd buy both. I like the hooded ones, as opposed to the ones with a helmet. I

was paranoid at first of the hood being in contact with my head, but I have three nylon outfits (one jacket and two coveralls), and two cotton, all with hoods, and have never been stung on the back of the head like I expected. My favorite jacket is the Ultra Breeze as it is mesh, sting proof and cool on a hot day. It's expensive and worth every penny.

Gloves

I would wear standard leather gloves and tuck them into the sleeves of the jacket. They will be easier to get on and off than the long ones and cheaper to buy.

Some kind of hive tool

Any little flat bar will work. One of my all time favorites is a very old light cleaver (the blade is about $1^1/_2$" wide and 6" long) that I sharpened on the end. I can pry a box apart or scrape things. It doesn't pull nails well and if the prying is really heavy I do worry about breaking it. If you're going to buy one, I really like the Italian Hive tool I got from Brushy Mt. It's got a lift hook on one end and is light and long has a lot of leverage. But I don't see it in their last catalog. My next favorite is the Thorne hive tool with a frame lifter and next is Maxant's Frame Lifter hive tool. But I do like the Italian one from Brushy Mt. better because the hook fits between the frames more easily.

A bee brush

You can buy one, or if you hunt or have birds you can use a large feather. It has to be a nice stiff quill to do any good. You will need to brush bees off from time to time. In order to harvest, in order to do other manipulations. Shaking can work sometimes, but sometimes you just need a brush. Like when the bees are all clustering on the edge of the hive you can brush them off before you set the next box on top.

Nice to Have Beekeeping Equipment:

These are nice, but not essential, you can do fine without them, but I don't think you will regret buying them.

Tool box

You can put your tools in a five gallon bucket, but if you want a really nice toolbox, Brushy Mt. has one that can double as a swarm box, has a place for a hive tool, a frame grip, a smoker, a frame perch and room inside for odds and ends. It makes a nice stool too. If you want to build your own, take a close look at the one Brushy Mt has and convert a nuc box.

Queen Catcher

The hair clip kinds are the nicest ones I've seen to pick up a queen without hurting her. You still have to be a little careful, but it is designed to not hurt her and to let the workers out. There are times you just need to know where she is while you rearrange things or do a split and then you can release her. This plus a marking tube and a paint pen and you can mark her too.

Queen Muff

I got one from Brushy Mt. You can catch the queen in the hair clip and put her in the muff and not worry about her flying off.

A frame nailing device

This device (Walter T. Kelly has these) is very nice to put wooden frames together. It holds 10 frames in place for you to nail them. It is a little tricky to figure out at first, but it's a real time saver and frustration saver.

A $^1/_4$" crown staple gun and compressor.

Everyone who owns a car needs a compressor anyway. The staple gun is under a $100. Walter Kelly has one that is the right size. It will shoot from $1^1/_2$" to $^5/_8$" staples (which I buy at the local lumber or hardware

The Practical Beekeeper 57

store). The 1″ are perfect for frames. The $1^1/_2$″ are perfect to put boxes together. The $^5/_8$″ are nice for when you don't want it to go through a $^3/_4$″ board and the $1^1/_4$″ are nice when you don't want to go through two $^3/_4$″ boards (like when you put a cleat on for a handle on a homemade box). Then you don't have to pre drill all those holes in the frames. I was a carpenter for years and am pretty good at nailing, but when doing frames I bend as many nails as I don't bend. Half of them are bent and pulled out when nailing by hand. But maybe my problem is I used to "one lick" a 16p nail and I don't have the finesse.

An Extractor

I would especially avoid buying a *new* extractor if you only have a few hives. If you find a good price on one, by all means pick it up, but buying a new one is a waste of money. Of course you can always keep your eye out for a bargain on a used one. I just crushed and strained and made cut comb for the first 26 years of my beekeeping. I finally bought a 9/18 radial when I started getting more hives. I'm glad I held out for a real extractor when I finally got one.

Avoid Gadgets

I would avoid all the gadgets out there as they will be superfluous and expensive. I like the Italian Hive tool from Brushy Mt. I would skip the frame holders and the frame grips and etc.

Useful Gadgets

Of the gadgets out there, I have enjoyed. I like the "Ready Date" nuc calendars as a way to keep track of the status of a hive. If you have outyards and haul a smoker around the smoker box from betterbee.com is a safety item worth having. You can put your smoker in it and not have to worry about catching your car on fire.

Getting Started

Now that we've covered equipment decisions, let's get started beekeeping

Recommended Beginning Beekeeping Sequence

I've thought about this and I'm sure a lot of people will disagree but I'm going to give my advice on how I would start beekeeping if I were a beginner doing it over again. This is what I wish I had done the first time.

First you have to decide how to get some bees. It's very difficult to get them from a tree or a neighbor's house when you really don't know anything about them. This is really an advanced undertaking. That said, I admit that is exactly what I did. I took them out of houses and trees and bought some queens. But I really didn't do so well at it and I got stung a lot. So all in all I don't think it was that good for the bees, although it was educational for me.

If you have local beekeepers you may be able to get a nuc or some frames of brood etc. The downside to this is they are probably on Deep frames (9 $^1/_4$" frames that go in a 9$^5/_8$" box). I'm not going to recommend deeps. They are also probably on large cell comb, and I'm going to recommend natural or small cell comb.

You can order package bees. I used to get them through the mail, but lately that has gotten more and more expensive. Most locations you can find a bee supply place that brings in a truck load of package bees in the spring. If you find a local bee club or association they will probably be able to advise you on this. Two packages would be a good start.

How Many Hives?

It is sound advice to get at least two hives. I think some beginning beekeepers don't get the *purpose* as they often want to experiment with two *different* kinds of hives, like a top bar hive and a Langstroth or a Langstroth eight frame medium and a Langstroth ten frame deep. But this defeats the purpose of having two hives. The main reason for having two hives is that the resource that is the hardest to come by and which is often needed to resolve issues of queenrightness, is frames of brood. But those frames of brood are not of much value if they are not interchangeable. If you really want a top bar hive and a Langstroth hive, then at least make them the same dimensions so the Langstroth frames are interchangeable with the top bars.

Package or Nuc?

Another issue new beekeepers often misunderstand is a nuc vs. a package. It comes down to this, if you want bees on some kind of frame or comb other than what they are on, buy a package. In other words if the nuc is Langstroth deeps on large cell comb and you want a top bar hive or a small cell hive or a medium hive, then it's not practical to buy a large cell deep nuc and expect to put them in a medium box or a top bar hive.

On the other hand if you can get a nuc on the cell size, or frame you want, a good nuc will have a two week head start on a package and if you can get *local* bees in a nuc, especially local bees that have already wintered as a nuc, you will have a great advantage as they will be acclimatized to your climate, and an over-wintered nuc always seem really to take off in the spring often surpassing even strong hives that have overwintered.

Don't get too sidetracked on this two week advantage though. It's great as I said, if they are on the cell size and frames you want, but if they are not, it is not only a lot of work to convert to another frame size, cell size, or hive type, but you will set them back at least that two weeks or more you would have gained with the nuc, in the process. So take that into account when deciding.

Race of Bees

Assuming you are going to buy a package of bees, the next decision is what race. I hate to not have an opinion, but I really haven't seen a race of honeybees I didn't like. Well, I did have some really mean ones once, but they were the same breed I had been raising for decades. I will recommend you get something that is not a hybrid and can be open bred by you with good results. Caucasian, Italian, Cordovan (Italian), Russian and Carniolans are all fine. Take your pick. If you can get locally raised queens, that's better but for those of us in the North, there are seldom packages available with Northern queens. You can requeen with some later after you get them started.

More Sequence

We covered choices in the previous chapter, now that we've made all these decisions, here's the order I'd get things in.

Observation Hive

    I know a lot of people will disagree with me, but I would buy an observation hive. They will say, correctly, that an observation hive takes more skill to run. But you will learn *so much* in just a few days of watching one, and so much in the first year of watching one, that I think they are invaluable. Even if they die or swarm, you should learn a lot. You used to be able to buy a nice four frame "Von Frisch" hive from Brushy Mt. I'm not sure if they still stock it as I haven't seen it in the catalog. It holds four medium frames (remember we want all the frames the same). You do have to make the hookup for the tube yourself but everything else is pretty much done for you. To hook up the tube I take a 1″ long 1″ diameter galvanized water pipe nipple and a $1^{1}/_{8}$″ hole saw (that goes in a drill to make a $1^{1}/_{8}$″ hole)

and glue a piece of pine in the end of the Von Frisch hive and drill the 1 $1/8$" hole and use some channel locks or a pipe wrench to screw in the pipe nipple. Get some $1^1/4$" tubing and attach it with a hose clamp. Cut a 1 x 4 to fit in under your window and another that fits under your storm window and drill a $1^3/8$" hole in both of those so that with the windows closed they line up. Thread the $1^1/4$" tubing (a sump pump kit works well) out through the window. I also added a screen molding behind the hinges and behind the door stop to increase the space between the glass by $1/4$". This works out perfectly. The $1^1/2$" space that it comes with works if the bees are drawing their own comb in the hive. But if you ever swap drawn frames from a hive it's too close and PermaComb or Honey Super Cell is also too close.

If you get on the bee forums there are people who make observation hives, often to your specifications.

Also I would put a very small screw or a staple in back and on the door in the frame rest area to hold the frame out at the correct space. I seem to always be carrying the hive back in from outside and jostle the frames and they slide to one side and mess up the beespace.

Make some frames (or wax dip some Perma-Comb) and put the small cell foundation in it. Put these in the observation hive. Cut some Black cloth so that doubled up and folded over the hive it covers both sides to the floor. This is a privacy curtain.

Nucleus Hive

When the bees outgrow the observation hive you will need somewhere to put them. If we are going with eight frame mediums we can just use one eight frame box for a nuc and all our equipment will be the same. If not, let's build or buy a medium nuc. Get a bottom and cover (or make them). This will make a good start for when they outgrow the observation hive. A nuc also gives you a place to keep a spare queen or make a small split and not leave them too much room. Put it together so it's ready before you get the bees. Now you wait for spring.

Putting Bees in the Observation Hive

Come spring put the bees in the observation hive. I assume this is a package, so you need to make sure they are well fed. Spray the screen lightly with sugar syrup waiting periodically and spraying again until they lose interest in eating it off of the screen wire. Take the bees and the observation hive outside near the en-trance to the observation hive. Cover the exit to the hive with a piece of cloth and a thick hair tie rubber band (they are easier to handle) Do the same with the outside entrance to the tube and the other end of the tube in the house. Lay the observation hive flat on its side on the ground and open the door. Put on your protective equipment. Pry open the lid to the box and carefully fish out the queen cage and set it aside. Now fish out the can and shake the bees off of it into the observation hive. Hit the box sharply on the ground to dislodge the cluster and then flip it upside down and pour the bees into the observation hive. Hit the box sharply on its side to knock the remaining bees to one end then dump them in. If there are still 20 bees or so in the box, don't worry about it. If there are hundreds of bees in the box, repeat the steps until there are only a few.

Spritz the queen lightly with some water so she won't be as likely to fly. Carefully pry the staple off of the queen cage, being careful not to open the screen and let out the queen. Put the queen cage over a cluster of the bees and holding the screen side down, open the screen and put the cage close to the bees watching for the queen to walk out. (Difficult, I know). If you didn't see her and you didn't see her fly off and you didn't see her go in, then we may have to keep an eye out for a while. Assuming she went in, use the smoker to drive the bees away from the door frame so they don't get squashed and close the door (squashing some stubborn and indecisive bees, but hopefully not too many.) Now brush all of the bees off of the outside of the hive and take it in the house. Holding the hose up to the pipe, pull off the cloth from both pieces and slide the hose on and clamp it (the clamp has to be on the hose before you do this.

You now have an observation hive. Fill a quart jar with 2:1 syrup (2 parts sugar to 1 part water) and feed them. Now go take the cloth off of the outside of the tube.

If you didn't see the queen go in, watch outside for any clusters of bees on the ground or bushes. If you see any, look carefully to see if there is a queen. If so, catch her with the hair clip catcher and put her at the tube entrance and see if she'll go in. If she doesn't, you may have to take the hive outside and do it all again, but probably you now have a queen in the hive.

If you bought two packages (recommended) then put the other in the nuc and buy your equipment for a hive and assemble it now.

Keep feeding them and watch them. Count the days until the queen starts to lay eggs. (usually at least three or four days but sometimes as long as two weeks) and how many days until they eggs hatch and how many days until you see capped brood and how many days until you see emergence. The hive will build slower at first but once bees start emerging the population will explode.

Making a Split into the Nucleus Box

When they have pretty much filled the hive with honey, brood and pollen you need to move three frames and the queen to the eight frame hive. Feed them and keep feeding the observation hive. Try to be sure that the frame you leave in the observation hive has eggs. Now you get to watch them raise a queen. By the time the queen in the observation hive is laying, all of the brood will have emerged. The Observation hive will be struggling again to get going, but the five frame nuc will quickly fill up and when four and one half frames of it are full, add the next box and order four medium boxes and enough frames for them and a screened bottom board and an inner and outer cover or a migratory cover. When the two eight frame boxes are full put the queen and all but two frames in the other hive. Make sure one of those frames has eggs and open brood and the other has pollen and honey. Put those two in one box with a top and bottom and let it raise a queen.

Now you have one hive, one nuc, and one observation hive (and if you bought a second package another hive). If you need a queen you can unite the nuc with the hive, or pull a frame of brood for the nuc to raise one or pull a frame of brood for the observation hive to raise one. You get to watch in detail what is going on with the bees in the observation hive. You can see pollen coming in, you can see nectar coming in, you can see when they are being robbed, you can see if they are having any problems. You can watch the queen lay. You can practice finding the queen without disturbing the hive.

Managing Growth

As the observation hive gets too strong you can pull frames out and put them in your regular hive to boost them. As the nuc gets too strong you can pull frames and put them in the regular hive. You can replace them with undrawn foundation. If you only want one hive, you have one and some spare parts to fix it. If you want another hive, just let the nuc grow and put it in a regular hive too. Then start another nuc from some frames from the observation hive so you have two hives a nuc and an observation hive.

Starting With More Hives

Of course if you wanted to start with more hives (a good idea actually), you could put a package in the observation hive and a package in the nuc or the hive at the same time. More redundancy lets you have resources to fall back on when they get into trouble. I wouldn't go with more than four hives to start off.

Foundation and Frames

What kind of foundation and frames should you buy? Obviously if there was a "right" answer, there would only be one kind of foundation and one kind of frames. The reason there is not is that beekeepers have different preferences and different philosophies and different experiences.

Let's get a little terminology out of the way. With wax, about the only thicknesses I see available now are "Medium Brood", "Surplus" and "Thin Surplus". "Medium Brood" does *not* mean it goes in medium frames. It means it is of medium thickness. Surplus is thin and "Thin Surplus" is even thinner. Surplus is intended for comb honey.

Brood foundation

The thing the bees like to build from the most is no foundation. Foundationless frames are the best accepted, and the most natural. They have many advantages from the Varroa control of smaller cells, to being able to cut out queen cells from a comb without worrying about hitting a wire or having plastic in the middle of the comb stop you.

The thing the bees like next is wax foundation. They can rework it to what they want. But the closer it is to what they want the better it will be accepted. I'd say, with unregressed ("normal") bees 5.1mm would be the best accepted, as that seems to be what they want to build. Dadant sells this. 4.9mm would be next and 5.4mm last. But I want the 4.9mm for the Varroa control aspect. So one aspect of foundation is the material (wax or plastic) and another is the size of the cell.

The other issue with wax foundation is reinforcement. DuraComb and DuraGilt have a smooth plastic core. This works well until the bees strip the wax off to use somewhere else or the wax moths eat down to the plastic. Then the bees won't rebuild on the plastic. Wires are often used in wax foundation. Some foundation comes with vertical wires in it and people use it as is. Some comes with none and some people wire it with horizontal wires. The wires slow down the process of the foundation sagging.

The material the bees seem to like the least and the beekeepers seem to like the most is plastic. The wax moths can't destroy the foundation (although they *can* destroy the comb). The bees can't rework the size very easily. Sizes of plastic vary from 5.4mm down to 4.95mm. It is available as sheets of plastic foundation or fully molded frames with foundation.

Fully "drawn" comb is also available in plastic. PermaComb (5.0mm equivalent cell size) is available in mediums and Honey Super Cell (4.9mm equivalent cell size) is available in deeps. Fully drawn, means the bees don't draw it out, it's already full thickness, and they just use it and cap it.

Foundation for supers

The fully drawn comb is certainly an advantage here (once the bees have accepted and used it) as the bees have only to store the nectar and don't have to build any comb. The wax moths can't touch it nor can the small hive beetles.

The various plastic frames and plastic foundation for supers are the same as the ones available for brood, with the additional use by some of drone comb (easier to extract) and Honey Super Cell's 6.0mm cell size with a fake egg in the bottom of the cell. The fake egg supposedly fools the queen so she won't lay in it. The 6.0mm also discourages the queen as it's not quite a drone size (6.6mm) nor a worker size (4.4mm to 5.4mm) so she doesn't like to lay in it.

For comb honey, there is surplus and thin surplus. This is so the comb honey will be easy to chew and not have a thick core in the middle. It is available from most manufactures. Walter T. Kelley has it in 7/11 which, again, is a size the queen doesn't like to lay in so you can forgo the excluder and not get brood in the supers.

Kinds of frames

There are different kinds of frames and many of the foundations were planned to be used in one or the other of them. You can usually adapt either way, but you may want to take this into account when ordering frames and when ordering foundations.

Top bars come in grooved, wedge, and split (split is available from Walter T. Kelley). The grooved are usually used with plastic or with a wax tube fastener. I prefer them to the wedge. I can attach a lot more foundation a lot more reliably (so that it doesn't fall out) with a wax tube fastener than a wedge. The wedge type has a cleat that breaks off and is nailed into the frame to hold the foundation. The split is usually used for comb honey. The foundation is just dropped down into the split onto a solid bottom bar and put in the hive without nailing at all.

Bottom bars come in split, grooved and solid. I prefer solid, as the wax moths won't get into them. But your foundation may not fit with a solid bottom bar (depending on what you buy). The split ones are not very strong and always seem to break the first time I try to clean them up and put new foundation in them.

Grooved are usually used for plastic so that the plastic foundation snaps into the frame. The other issue is the exact size of the foundation you are using. Some is cut to go all the way to the bottom with split frames. Some is cut to fit in the groove. Walter T. Kelley seems to be the only supplier who carefully maps what fits in what in their catalog.

Plastic one-piece frames. These eliminate all the issues, other than acceptance and cutting out queen cells. No frames to build. The foundation obviously fits since it's already in there. If you buy Mann Lake PF-120s (medium depth) or PF-100s (deep depth) they are 4.95mm cell size so you get the advantage of small cell. They are cheap (in large lots they are a little over $1 each last I saw). There is no wiring to do and they are well accepted by the bees.

Locating hives?

"Where should I put my hive?" The problem is there is neither a simple answer nor a perfect location. But in a list of decreasing importance I would pick these criteria with a willingness to sacrifice the less important ones altogether if they don't work out:

Safety

It's essential to have the hive where they are not a threat to animals who are chained or penned up and can't flee if they are attacked, or where they are likely to be a threat to passersby who don't know there are hives there. If the hive is going to be close to a path that people walk you need to have a fence or something to get the bees up over the people's heads. For the safety of the bees they should be where cattle won't rub on them and knock them over, horses won't knock them over and bears can't get to them.

Convenient access

It's essential to have the hive where the beekeeper can drive right up to it. Carrying full supers that could weigh from 90 pounds (deep) down to 48 pounds (eight frame medium) any distance is too much work. The same reason applies for bringing beekeeping equipment and feed to the hives. You may have to feed as much as 50 pounds or more of syrup to each hive and carrying it any distance is not practical. Also you will learn a lot more about bees with a hive in your backyard than a hive 20 miles away at a friend's house. Also a yard a mile or two from home will get much better care than one 60 miles from home.

Good forage

If you have a lot of options, then go for a place with lots of forage. Sweet clover, alfalfa being grown for seed, tulip poplars etc. can make the difference between bumper crops of 200 pounds or more of honey per hive and barely scraping a living. But keep in mind the bees will not only be foraging the space you own, they will be foraging the 8,000 acres around the hives.

Not in your way

I think it's important the hive does not interfere with anyone's life much. In other words, don't put it right next to a well used path where, in a dearth and in a bad mood, the bees may harass or sting someone or anywhere else where you are likely to wish they weren't there.

Full sun

I find hives in full sun have fewer problems with diseases and pests and make more honey. All things being equal, I'd go for full sun. The only advantage to putting them in the shade is that you get to work them in the shade, or it might help meet one of the other more important criteria.

If you live in a very hot climate, mid afternoon shade might be a nice to have, but I wouldn't lose sleep over it unless you have a top bar hive; then I would go for shade to prevent comb collapse.

Not in a low-lying area

I don't care if they are somewhere in the middle between low and high, but I'd rather not have them where the dew and the fog and the cold settle and I really don't want them where I have to move them if there's a threat of a flood.

Out of the wind

It's nice to have them where the cold winter wind doesn't blow on them so hard and the wind is less likely to blow them over or blow off the lids. This isn't my number one requirement, but if a place is available that has a windbreak it's nice. This usually precludes putting them at the very top of a hill.

Water

Bees need water. One of the issues is providing it. Another is to have it more attractive than the neighbor's hot tub. To accomplish this you need to understand that bees are attracted to water because of several things:

- Smell. They can recruit bees to a source that has odor. Chlorine has odor. So does sewage.
- Warmth. Warm water can be taken on even moderately chilly days. Cold water cannot because when the bees get chilled they can't fly home.
- Reliability. Bees prefer a reliable source.
- Accessibility. Bees need to be able to get to the water without falling in. A horse tank or bucket with no floats does not work well. A creek bank provides such access as they can land on the bank and walk up to the water. A barrel or bucket does not unless you provide ladders or floats or both. I use a bucket of water full of old sticks. The bees can land on the stick and climb down to the water.

Conclusion

In the end, bees are very adaptable, so make sure it's convenient for you, and if it's not too hard to provide, try to meet some of the other criteria. It's doubtful you'll have a place that meets all of the criteria listed above.

Installing Packages

It occurs to me listening to all of the newbees on the bee forums and watching the U-Tube videos of inexperienced people doing their first installs and listening to the experts give advice at new beekeeper classes etc., that there is a lot of very bad advice out there. Sometimes it's just that a beginner doesn't know what a happy medium of something is, but all in all, I think it's just bad advice. So here's my take on a lot of that advice of what to do and not to do:

Not to do:

Don't spray them with syrup

Certainly if you insist on doing this, don't spray them much and don't use thick syrup. 2 parts water to 1 part sugar is plenty. Personally I would not and do not spray them at all. If you have to feed them because you can't get them installed, just spray a little on the screen and wait for them to clean it up. Repeat until they don't take it. But actually I think it's a better plan to refill the can with syrup. Pull it out (of course the bees can now get out so put a board or something over the hole). If you have the kind of can that has a round hole with a rubber grommet holding in a piece of cloth, pop this out and pour in the syrup. Replace the grommet and cloth and then replace the can. If there are just the small holes, then put a hole just big enough for the syrup to run in and fill it full of syrup. Then plug the hole with some softened beeswax. Check for leaks and put the can back.

Why? I've seen many drowned sticky bees from leaky cans or spraying of bees or worse, from over-heated bees that regurgitate their honey stomachs as a reflex to cool them off. I don't want to see any more drowned bees. I watched a U-Tube video the other day of someone knocking the bees to the bottom (which is fine if you're about to dump them into the hive) soaking them (literally) with syrup, turning the box around and soaking them some more from the other side, then after messing with the hive a bit, soaking them again. I doubt if half of them lived.

I've never seen bees die from *not* spraying them with syrup.

Don't leave them in the shipping box

Don't put them in the hive in the shipping box in order to avoid dumping them out — especially if the box is on top of the top bars with an empty box on top. This is just asking for problems. Assuming you put the queen cage somewhere in the hive; the bees will cluster on the inner cover or cover and then draw combs in the empty box. Bees always prefer their own comb to drawing on foundation and will take every opportunity you give them to do so. Don't give them that opportunity. Bees are not hard to dump out of a box. Yes, this is one of those few things where gentleness and grace are not helpful, but that does not make it hard on the bees or upsetting to the bees. You may as well get used to the idea as someday you'll be shaking a swarm into a box instead of a swarm out of a box. If you really insist on letting them leave the box on their own, then put an empty deep (or medium or whatever) on the *bottom* and put the box in there and then put a box with frames on *top* of that. This takes advantage of the fact that the bees will try to cluster at the top and hang down from there. So hopefully that will be the inner cover and not the bottom bars. Make sure you remove the shipping box and the empty box *the next day*. Not four days later. Not five days later—t*he next day*. Otherwise you risk them building comb in the empty space.

Don't hang the queen between the frames

This almost always results in an extra comb between those two frames drawn on the queen cage. Release the queen and you won't have to worry about the messed up combs. This is even more important in a foundationless scenario such as a top bar hive or foundationless frames as one messed up between the frames comb will result in a repeat of the error the rest of the way across. Dump the bees in. Let them settle a bit. To keep the queen from flying, pull the cork from the non candy end (where she can get out now) and, while holding your thumb over the hole, lay the cage on the bottom and leave it. Put the frames back in and the lid on and walk away. Don't try to release her onto the top bars. Release her down on the bottom board.

One of the issues seems to be that people think that either they will abscond or they will kill the queen. In my experience leaving her caged does not seem to resolve these issues. If they want to leave they usually move to the hive next door anyway and abandon the queen. If you release the queen it also won't stop this from happening, but it also won't cause it. I've not had a problem with a package killing the queen. A bunch of confused bees have been shaken together from many hives and in the confusion they are just happy to find a queen. If they do kill the queen it is almost always because there is already one loose in the package that got shaken in. The bees prefer this queen because they have contact with her.

Don't use an excluder as an includer too long

Don't use an excluder as an includer (to keep the queen *in*) after there is open brood in the hive. I wouldn't use it at all, but there is no point in it after there is open brood and it will keep the drones from being able to fly.

Don't spray the queen with syrup

It will make a mess. Yes, it will probably keep her from flying, but it could also do her harm. I know some think it doesn't but they apparently have not seen a half dead sticky queen before. I've seen plenty. I don't spray her with anything, but if you insist, just use water or at most 2 parts water to 1 part sugar.

Don't install bees without protective equipment

You have enough to worry about without worrying about them stinging you as well.

Don't smoke a package

They are already in a docile mood and they need the pheromones to get organized, find the queen etc. There is no need to interfere with these pheromones as smoking will do little to nothing to calm a swarm or a package anyway.

Don't postpone

Don't postpone installing them because it's a little drizzly or chilly. Unless it's like 10º F or less I would install them and consider it an advantage that they won't want to fly and they will settle in better anyway.

Just make sure you have food for them so they don't starve. Capped honey is best. Dry sugar that has been sprayed with enough water to get it damp will do.

Don't feed in a way that makes excessive space

A package is a comb building team. They are looking to build comb everywhere they can. Don't give them space to build it places where they shouldn't. This includes putting empty boxes on top that they have access to, or a spacer for a baggie feeder etc. A frame feeder, a jar over the inner cover with duct tape covering any access or something similar is good. A bottom board feeder is good. Baggie feeders on the bottom board are good *if* you put the bees in first and the baggie feeders on after the bees are off of the bottom.

Don't leave frames out

Ever. Not even for a few minutes. Often you intend to leave them out for a few minutes and forget to come back. When you close a hive up there should always be a full complement of frames in the box, or in the case of a top bar hive, a full complement of bars. Even if you use a follower to temporarily limit the space, fill the empty space with frames or bars. You never know when the bees will find their way over there.

Don't dump bees on top of a baggie feeder

They will get covered in syrup as it all gets squished out by the weight of the bees falling on the baggie.

Don't close up a newly hived package

Let them fly and breathe and get oriented.

Don't leave empty queen cages around

The bees will cluster on them and act like a swarm thinking the cage is a queen because it still smells like one.

Don't let messed up comb lead to more messed up comb

If you have foundationless or a top bar hive this is even more critical. With foundation you get a sort of clean slate every frame as there is another wall of foundation to start from. Still I would try to straighten out any messes quickly. Bees build parallel combs, so with foundationless one bad comb just leads to another. By the same token one good comb leads to another as well. The sooner you make sure the last comb from which the "next" is being built is straight and centered; the better off you will be because the next comb will be parallel to that one. If you have a top bar hive, make sure you have some frames built that you can tie combs into if they get crooked or fall off. That way you can always get at least the last one in the row straight again or, better yet, all of them straight. Especially with foundationless, I would check soon after installation and make sure they are off to the correct start, meaning the combs are in the frames and lined up correctly. The sooner you make sure, the better off you'll be.

If you're using foundation and the bees build fins off of the foundation or parallel combs where there is a gap you can't get to, scrape this off before it has open brood in it. The wax isn't nearly the investment that open brood is. Keep the hive clean of this messed up comb or it will haunt you for a long time to come. With plastic foundation you can just scrape it to the plastic. With wax foundation you'll need more finesse.

Don't destroy supersedure cells

Packages often build supersedure cells and they often tear them back down after a few days, but you tearing them down will risk them ending up queenless. Sometimes there is something wrong with the queen that you don't know. Assuming that the bees are mistaken and you are correct about the quality of the queen is, in my experience, a bad bet.

Don't panic if the queen in the cage is dead

Don't panic and assume they are queenless if the queen in the cage is dead when you get it. Odds are there is a queen loose in the package. Still I would contact the supplier just in case, but meantime install them and come back and check them before you install that new queen. You may just be sentencing her to her doom.

Don't freak out if the queen doesn't lay right away

Some will lay as soon as there is comb $1/4''$ deep in the hive. Some take as long as two weeks to start to lay. If they aren't laying in two weeks they probably aren't going to and it's time to freak out.

Don't freak out if one hive is doing better than the other

There are many contributing factors. If they have eggs and brood they are probably doing fine.

Don't get just one hive

Get at least two. You'll have resources then to deal with issues that will come up.

Don't feed constantly

Don't just keep feeding figuring they will stop taking it when they don't need it. I've seen packages that swarmed when they hadn't even finished the first box because they backfilled it all with syrup. Feed until you see some capped stores. This is the sign that the bees have put some of it in "long term storage" meaning they consider it a surplus. If there is a nectar flow at that point, I would stop feeding.

Don't mess with them everyday

They may abscond if you mess with them too often.

Don't leave them on their own for too long

You'll miss the opportunity to learn and you may miss that things are not going correctly. I would check on them within three or four days for the first time and then wait at least that long between visits and try not to go through everything. Just get a general idea how things are going.

Don't smoke them too much

Don't smoke them too much when working them after the install. The most common smoking mistakes:

- People have the smoker too hot and burn the bees with the flame thrower they are wielding
- People use far too much smoke causing a general panic instead of simply interfering with the alarm pheromone. One puff in the door is enough. Another on the top if they look excited is ok and after that having it lit and setting nearby is usually sufficient.
- People don't light the smoker because they think smoke upsets the bees, probably because of one of the above reasons.
- People blow the smoke in and immediately open the hive. If you wait a minute the reaction will be completely different. If you're doing something not too time consuming, like filling frame feeders or something, it's a good plan to smoke the next hive before you open this one. That way the minute will be up when you open that one.
- People don't smoke because they have the idea that it is either bad for the bees or somehow unnatural. Their exposure is only a puff or two once every week or two. People have been smoking bees for at least 8,000 years that we have documented for one very good reason. Nothing works better at calming them.

Things to do:

Always install them in the minimum amount of space

It takes heat and humidity to raise brood and make wax. Always install them in the minimum amount of space that is large enough and is convenient for you to provide. In other words, if you have a five frame nuc box, that's excellent. If you don't, then use a single box. Yes a single five frame medium box is large enough if you don't have drawn comb in it. An eight frame medium box is large enough if it has drawn comb. While there is nothing wrong, per se, with putting them in more space, in a Northern climate, especially, it is a lot of work for them and they take off much better in a smaller space. While I probably wouldn't *buy* a five frame nuc just for this, I would use it if I had it.

Have your equipment ready

Have your equipment ready before the bees arrive. Have the location picked and the equipment there. Have your protective equipment too.

Wear your protective equipment

You have enough to worry about without thinking about getting stung.

How to install:

When you have everything there, bees, equipment etc, then pull out four or five frames, pull out the can and the queen, slam the box on the ground to knock the bees loose and pour them out like thick oil, or like getting a pick out of a guitar. Tip the box back and forth as needed and when no more will pour slam it again to knock them loose and pour some more. When you are down to ten or twenty bees, set the package down. Pull the cork on a non candy end (if there is candy) of the queen cage and hold your finger over the hole and set it on the bottom board and let go. Gently set the frames in. Do not push them down on the bees on the bottom. Let the bees move and the frames will settle on their own.

If you release the queen (trickier to make sure she doesn't fly) then do *not* leave the cage. Shake all the bees off of it and put it in a pocket and take it in the house when you are done. Otherwise the bees will cluster on the cage and you'll end up with a queenless swarm on the cage.

Frames tightly together

For some reason this seems to be ignored in the books and causes no end of problems for beekeepers. The frames should be tightly together in the center. A full complement of them (10 for a 10 frame box). If you leave excess space, the bees are likely to do something funky between, like an extra comb or one out from the face of the comb or fins off of the face of the comb. Your best prevention for this kind of "creative" construction is to push them tightly together. Better yet shave them down to $1^1/_4$" wide and put an extra frame in tightly together.

Do feed them

A package will go through a lot of feed especially when they have no comb and no stores. Feed them until you start seeing capped honey or they start to backfill the brood nest. Do check on them to make sure things are going correctly. Better to catch things sooner than later, especially things like misdrawn comb.

Enemies of the Bees

Traditional Enemies of Bees

Traditionally bees have had enemies; pests, predators and opportunists. Some are as large as a Bear and some are as small as a virus.

Bears.

Ursa. Bears are not a problem for me. Some people live where there are bears and they are their biggest problem. All kinds of bears love to eat bee larvae and they don't mind honey too much either. Symptoms that you have a bear problem: Hives all tipped over and large chunks of the brood nest eaten. Sometimes vandals will tip over hives, but human ones don't usually eat the larvae. The only solutions I've heard of for bears are very strong electric fences with alternating ground wires on the fence (so they are sure to get grounded) and bait on the fence (bacon is popular) so that the bear gets its tender mouth parts on the fence. This seems to work most of the time. Some people put the hives up on a platform too high for the bears, but it is difficult to haul honey down from the platform and move boxes up. Of course sometimes the only way to stop a bear is to kill it and eat it. However this leaves a vacuum and another bear usually soon fills it. The legalities, difficulties and dangers of that method are best left to a hunting magazine.

Bees Robbing

BLUF: *if you have robbing you need to stop it* immediately! *Damage progresses quickly and can devastate a hive. Just make sure they are robbing and not orienting first, then if it's robbing, do something drastic. Close off the hive, cover it with wet cloth. Open all the strong hives to make the strong hives stay home and guard their own hives. But do something even if it's as simple as closing off the hive with screen wire completely. Then you can assess what you want to do to let them fly (small entrance, robber screen etc.). Bottom line, you cannot let robbing continue. You need to stop it now.*

Sometimes during a dearth the strong hives will rob the weak ones. Italians are particularly bad about this. Feeding seems to make this worse or sometimes set it off. Prevention is best. When you see that a dearth is setting in, reduce the entrances on all the hives. This will slow them all down some. But you need to have an eye on them to see that the dearth is over and open them back up during a flow.

I've noticed that queenless hives get robbed much more often than queenright hives. I had always thought it was because the robbers kill the queen, and they probably do, but when I make a nuc queenless in the fall just before I combine them with another nuc they seem to get robbed almost immediately.

One issue is being sure they are being robbed. Sometimes people mistake an afternoon orientation flight with robbing. Every warm, sunny afternoon during brood rearing you'll see young bees orienting. They will hover and fly around the hive. This is easily mistaken for robbers who also hover around a hive. But with practice you'll learn what young bees look like doing this. Young bees are fuzzy. Young bees are calm compared to robbers. Look at the entrance. Robbers are in a frenzy. Local bees might have a traffic jam at the entrance but they will still be orderly. Wrestling at the

entrance is pretty much a give away, but lack of fighting at the entrance does not prove they are not being
robbed; it just proves they have overcome the guard
bees. One *sure* way to tell if they are being robbed is to
wait for dark and close the entrance. Any bees in the
morning who show up trying to get in are probably
robbers — especially if there are a lot of them.

Inside view of robber screen.

Outside of robber screen

If you already have robbing occurring, here are some
ways to stop it. A really weak hive can be closed up

with some #8 hardware cloth for a day or two. The robbers can't get in and eventually get tired of trying. It helps if you can feed and water them. A little bit of pollen and a few drops of water will get a small nuc by. More will be required if there are more bees. After you open back up be sure to reduce the entrance. If you can feed, water and ventilate for 72 hours, you can close them up when they are full of robbers and force the robbers to join the hive. Another variety of confining them is to stop up the entrance with grass. The bees will eventually remove it, but hopefully the robbers will give up before then.

A "robber screen" can be built from scratch or you can use a screen door from Brushy Mt. (they seem to have modified theirs to work as a robber screen now). It is a screen that covers the area around the door and has an opening in the top (you will have to make the whole affair). This forces the robbers to turn a couple of corners to find their way in. Since they seem to go by smell this confuses them. It also stops skunks.

Vicks Vaporub around the entrance will also confuse the robbers because they can't smell the hive. It does not confuse the bees that live there because they remember how they got out.

A weak hive will sometimes get totally robbed out so there is not a drop of honey left. They will quickly starve. If you can't control the robbing it's better to combine some of the weak hives than let them get robbed out and starve. If you only have one strong and one weak, you can steal some emerging brood from the strong hive to boost the weak hive and shake off some nurse bees (the ones on the open brood) from the strong hive in the weak hive. Or you can just combine the weak with the strong. It's better than all the fighting and starving.

Skunks

Mephitis mephitis and other varieties. Skunks are a common predator of bees all over North America. Symptoms are very angry hives, scratches on the front of the hives, little soggy piles of dead bees on the ground near the hives that have had the juice sucked out of them. Many solutions work fairly well. Putting the hives up higher or having a top entrance, carpet tack strips on the landing board, chicken wire on the landing board, robber screens, trapping, poisoning and shooting. I have really only done the shooting and screen doors, and ended up doing top entrances. But many swear by the other solutions. A raw egg in the shell with the end removed and three crushed aspirin in it with the other end of the egg buried in the ground in front of the hive(s) being harassed is one solution I've heard of that I would probably have tried if the top entrances had failed. Other poisons worry me because of my dog, chickens and horses.

Opossums

Didelphis marsupialis. Pretty much same problems and solutions as the skunks.

Mice

Genus Mus. Many species and varieties. Also shrews (Cryptotis parva). Mostly these are a problem during winter when the bees are clustered and the mice move in. Using #4 hardware cloth ($^{1}/_{4}''$ squares) over the entrances will let the bees in and out and not the mice. Or use only an upper entrance so the mice can't get in.

Wax moths

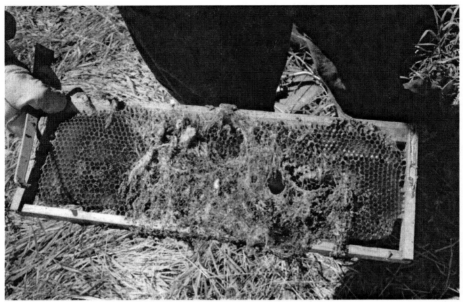

(Photo by Theresa Cassiday)

Galleria mellonella (greater) and Achroia grisella (lesser) wax moths are really opportunists. They take advantage of a weak hive and live on pollen, honey and burrow through the wax. They leave a trail of webs and feces. Sometimes they are hard to spot because they try to hide from the bees. They burrow down the mid rib (mostly in the brood chamber but sometimes in the supers) and they burrow in the grooves in the frames. This seems to preoccupy a lot of beekeepers and be the cause of a lot of chemical contamination in the hive, so let's address it here.

Climate

First, understand that this is a very climate de-pendent issue. In a climate where you seldom if ever have a hard freeze the wax moths may live year round which will be an entirely different scenario than in a climate where you get hard freezes and a long winter. I will share what I do and how that works, but keep in mind you will need to adjust this to your climate and your situation and indeed, if you live where the wax moths never die from cold the method I use will not work at all and a different method will have to be used.

Cause of Wax Moth Infestation

First, let's talk a bit about the moths. Galleria mellonella (greater wax moth) and Achroia grisella (lesser wax moth). Both will invade unguarded comb during the season when they are active. They prefer comb with pollen in it and as a second choice comb with cocoons in it, but they will even live on pure wax with nothing in it. Most of my wax moth issues are when a walk away split fails to raise a queen and the hive dies, or a mating nuc dwindles too much to guard the comb well enough. I really don't have any other wax moth issues, but in the past have when I have made some drastic mistakes.

Beekeeping Mistakes

One year, based on someone else's shared experience, I left the boxes wet and put them in my basement. The wax moths not only destroyed all of those combs but so infested my house that I have never gotten rid of them. There have been wax moths flying around my house every since and that was in 2001. Never put supers, especially wet ones, in a warm place. Especially when you have the option to put them outside where they will freeze and the moths will die. That they have to have brood comb is a myth. They prefer brood comb, yes, but they do not require it.

Wax Moth Control

My current method is this. I wait until late to harvest. The reasons for this are that I can better assess what I should leave for winter, thus saving feeding nearly as often, I save harvesting and then feeding, which is less work. I don't have to chase the bees out of the supers as I merely have to wait for a cold day when the bees are hunkered down and pull the supers which are empty of bees. After harvest I can put the wets on the hives and wait for some warm days for them to clean them up and after they are done, pull them off and stack them with no fear of wax moths as the weather is now cold and there are no moths around. If I want to harvest early, then I'd put the wet boxes back on and not pull them off until after a hard freeze.

The moths, in my part of the country, don't really get going until about late July or August and I try to have all of that drawn comb back on the hives by mid June at the latest, where the bees can guard them. So, I have no moths in the combs during the honey season (June through September), because they are guarded by the bees. I have no moths in the combs from October to May because we get freezing weather now and then which kills the moths and the moth eggs. I have none from May to June because they moths haven't built back up from winter yet.

Infested Bee Colony

What to do with an infested colony. The reason a colony gets infested is that it is weak. Prevention is not to give them more territory than they can guard, in other words, don't leave a lot of drawn comb on a hive that is small and struggling. Once they are infested, the solution is to reduce them down to just the space the cluster of bees can cover. Remove all the rest of the comb. If you have a freezer, freeze it to kill the moths, or if it's too far gone, let the moths finish cleaning it up. If they get to go to the logical end they will turn all the comb into webs that just fall out of the frames or off of plastic foundation. If it's only got a tunnel or two in it, freezing is a way to save the comb. I usually only have problems with colonies that have died out because they have gone queenless or gotten robbed out. In my management style, I find another nice thing about foundationless frames is you can give them to a hive and it's just empty space for future expansion, not all that surface area to guard from the moths as you have with wax foundation. Also nice in bait hives because the bees will build in the frames but you don't have wax moths tearing up the foundation.

Bt aka Bacillus thuringiensis

Some people use Bt (Bacillus thuringiensis) as either Certan or Xentari, on the combs. It will kill the moth larvae and seems to have no ill effects on the bees and studies have supported this view. It can be sprayed on infested combs even with the bees on them to clear up the infestation. It can be sprayed on foundation before putting it in the hive. It can be sprayed on combs before storing them. I simply haven't had the time to do this in recent years now, but, as I say, my management seems to keep them under control except in failing hives. But it would probably help in the failing hives if I had it on the combs ahead of time. Certan used to be approved for use on wax moths in the US but the certification ran out and there was no money in renewing it, so it's no longer labeled that way in the US, but is available labeled for that use from Canada and available labeled for use against moth larvae (but not wax moth per se) in the US as Xentari.

Tropical Wax Moth Control

What I would do if I lived in a more tropical area where moths don't get killed by winter: I would put empty combs on top of strong hives so they can guard them. This isn't a good plan in a temperate climate.

What not to do for Wax Moths

What I would not do, and is at the top of my list of things not to do, is use moth balls, particularly the Naphthalene ones. Slightly better, and on the FDA list as approved, is PDB (Para Dichlorobenzene). But both of these are carcinogens and I have no use for such things in my food supply, and beehives are part of my food supply.

Hating Wax Moths

I have given up hating wax moths, which is an easy thing to do when you see them destroy the combs the bees have worked at building. The wax moths are just part of the ecosystem of the beehive. They do their work and it is probably useful work. They get rid of old comb that might have disease lurking in the cocoons. If you really do hate them and want to keep them even more under control, which I have given up on, you can make traps. Basically a two liter bottle with small holes in the sides and a mixture of vinegar, banana peel and syrup inside seems to work well. It also catches a lot of yellow jackets. The moths fly in the holes in the sides, drink, try to fly up and get trapped.

Nosema

Caused by a fungus (used to be classified as a protozoan) called Nosema apis. Nosema is present all the times and is really an opportunistic disease. The common chemical solution (which I don't use) was

Fumidil which has been recently renamed Fumagilin-B. In my opinion the best prevention is to make sure your hive is healthy and not stressed and feed honey. Research has shown that feeding honey, especially dark honey, for winter feed decreases the incidence of Nosema. Also research done in Russia in the 70's has shown that natural spacing ($1^1/_4$" or 32mm instead of the standard $1^3/_8$" or 35mm) reduces the incidence of Nosema.

In my opinion moisture in the hive in winter, long confinement, any kind of stress and feeding sugar syrup increases the incidence. By all means, feed sugar syrup if you don't have honey and it means helping a struggling package or nuc or split. By all means, if you don't have honey, feed sugar syrup in the fall rather than let them starve, but, in my opinion if you can, try to leave honey on for their winter stores.

If you want a solution and don't want to use "chemicals" but want to use essential oils and such, thymol or lemongrass oil in syrup is an effective treatment. But keep in mind that these will kill many of the beneficial microbes in the hive as well.

Symptoms are a swollen white gut (if you disassemble a bee) and dysentery. Don't rely simply on dysentery. All confined bees get dysentery. Sometimes bees get into rotting fruit or other things that give them dysentery but it may not be Nosema. The only accurate diagnosis is to find the Nosema organism under a microscope.

If you want to get a grasp of how necessary (or not) it is to give preventative treatments for Nosema, I will point out a few things that may help clarify this for you. First, realize that many beekeepers have never treated for it, including me. Not only are there many beekeepers who don't want to put antibiotics in their hives, but in fact many beekeepers in the world are prohibited from using Fumidil by law. I am certainly not the only person who thinks it's a bad idea to put Fumidil in your hive. The European Union has banned its use in beekeeping. So we know they aren't using it legally anyway. Their reason? It is suspected of causing birth defects. Fumagillin can block blood vessel formation by binding to an enzyme called methionine aminopepti-dase. Targeted gene disruption of methionine amino-peptidase 2 results in an embryonic gastrulation defect and endothelial cell growth arrest. What do they use for treatment in the EU? Thymol syrup.

So why would you want to avoid Fumidil?

Just how dangerous is Fumidil to your hive? It's hard to say exactly, but of all the chemicals people put in hives, it's probably one of the least dangerous. It does break down quickly. It doesn't appear to have a lot of downsides on the surface anyway. But if you're of the Organic kind of philosophy you're still thinking, why do I want to add antibiotics to my hive? I certainly don't want it in my honey and, in my view; anything that goes in the hive can end up in the honey. Bees move things all the time. Every book I've seen on comb honey talks about the bees moving honey from the brood chamber up to the comb honey supers during a cut-down split. Having an area of the hive that is the only part there when chemicals are applied is a nice idea, but it's a lot like a no-peeing section in a swimming pool.

Microbial Balance

What do antibiotics do to the natural balance of a natural system? Experience with antibiotics would say that they upset the natural flora of any system. They kill off a lot of things that perhaps should be there along with what shouldn't leaving a vacuum to be filled by whatever can flourish. Probiotics have become a big thing in people and horses and other animals now, mostly because we use antibiotics all the time and upset the normal flora of our digestive system. Are there beneficial microorganisms living in bees and beehives? Are they affected by Fumidil? Yes, it's un-scientific of me to assume there are without some study to support it, but my experience says all natural systems are very complex all the way down to the microscopic level. I don't want to risk upsetting that balance.

Then there is the reason that it is outlawed in most of the world, which is that it causes a very specific birth defect in mammals.

Propping up weak bees

Yes, those with the Scientific philosophy will find that statement offensive. But I know of no better way to say it. Creating a system of keeping bees that is held together by antibiotics and pesticides; that perpetuate bees that cannot live without constant intervention; is, in my organic view of beekeeping, counterproductive. We just continue to breed bees who can't live without us. Perhaps some people get some satisfaction of being needed by their bees. I don't know. But I would prefer to have bees who can and do take care of themselves.

What other non-organic practices may contribute to Nosema?

Encouraging Nosema?

While the non-organic group tends to want to be-lieve that feeding sugar instead of leaving honey will prevent Nosema, I have seen no evidence of this. Honey may have more solids and may cause more dysentery, but while dysentery is a symptom of Nosema, it is neither the cause nor is it evidence of Nosema. In other words, just because they have dysentery does not mean they have Nosema.

Many of the Honey Bee's enemies, such as Nosema, Chalkbrood, EFB, and Varroa all thrive and reproduce better at the pH of sugar syrup and don't reproduce well at the pH of honey. This, however, seems to be universally ignored in the beekeeping world. The prevailing theory on how Oxalic acid trickling works is that the bee's hemolymph becomes too acidic for the Varroa and they die, while the bees do not. So how is it helpful to feed the bees something that has a pH in the range that most of their enemies, including Nosema, thrive, rather than leave them honey that is in the pH range where most of their enemies fail?

Bottom Line

The bottom line is this. You have to make up your mind what your risks are. What you are willing to put in your hives and therefore into your honey. How you want to keep bees. How much you trust a natural system or how much you want to strive for "better living through chemistry."

Stonebrood

This is caused by a number of fungi Aspergillus fumigatus and Aspergillus flavus. Extracts from this fungus are used to make Fumagillin used to treat Nosema. Larvae and pupae are susceptible. It causes mummification of the affected brood. Mummies are hard and solid, not sponge-like as with chalkbrood.

Infected brood become covered with a powdery green growth of fungal spores. The majority of spores are found near the head of the affected brood. The main cause is too much moisture in the hive. Add some ventilation. Prop open the inner cover or open up the SBB. Treatment is not recommended. It will clear up on its own.

Chalkbrood

This is caused by a fungus Ascosphaera apis. It arrived in the US in about 1968. The main causes are too much moisture in the hive, chilled brood and genetics. Add some (but not too much) ventilation. Prop open the inner cover or open up the SBB. If you find white pellets in front of the hive that kind of look like small corn kernels, you probably have chalkbrood. Putting the hive in full sun and adding more ventilation usually clears this up. Honey instead of syrup may contribute to clearing this up, since sugar syrup is much more alkali (higher pH) than honey.

"Lower pH values (equivalent to those found in honey, pollen, and brood food) drastically reduced enlargement and germ-tube production. Ascosphaera apis appears to be a pathogen highly specialized for life in honeybee larvae."—Author. Dept. Biological Sci., Plymouth Polytechnic, Drake Circus, Plymouth PL4 8AA, Devon, UK. Library code: Bb. Language: En. Apicultural Abstracts from IBRA: 4101024

Chalkbrood

Hygienic queens will also contribute to clearing this up. Hygienic bees will remove larvae before they fungus has created spores. The upside of chalkbrood is it prevents EFB.

European Foulbrood (EFB)

Caused by a bacteria. It used to be called Streptococcus pluton but has now been renamed Melissococcus pluton. European Foul Brood is a brood disease. With EFB the larvae turn brown and their trachea is even darker brown. Don't confuse this with larvae being fed dark honey. It's not just the food that is brown. Look for the trachea. When it's worse, the brood will be dead and maybe black and maybe sunk cappings, but usually the brood dies before they are capped. The cappings in the brood nest will be scattered, not solid, because they have been removing the dead larvae. To differentiate this from AFB use a stick and poke a diseased larvae and pull it out. The AFB will "string" two or three inches. This is stress related and removing the stress is best. You could also, as in any brood disease, break the brood cycle by caging the queen or even removing her altogether and let them raise a new one. By the time the new one has hatched, mated and started laying all of the old brood will have emerged or died. If you want to use chemicals, it can be treated with Terramycin. Streptomycin is actually more effective but is not approved by the FDA and the EPA.

American Foulbrood (AFB)

Caused by a spore forming bacteria. It used to be called Bacillus larvae but has recently been renamed Paenibacillus larvae. With American Foul Brood the larvae usually dies after it is capped, but it looks sick before. The brood pattern will be spotty. Cappings will be sunken and sometimes pierced. Recently dead larvae will string when poked with a matchstick. The smell is rotten and distinctive. Older dead larvae turn to a scale that the bees cannot remove.

Holts milk test:

The Hive and The Honey Bee. "Extensively Revised in 1975" edition. Page 623.

"The Holst milk test: The Holst milk test was designed to identify enzymes produced by B. larvae when speculating (Host 1946). A scale or toothpick smear is swirled gently into a tube containing 3-4 milliliters of 1 per cent powdered skim milk and incubated at body temperature. If the spores of B. larvae are present, the cloudy suspension will clear in 10-20 minutes. Scales from EFB or sacbrood are negative in this test."

Test kits are available from several of the bee suppliers. Free testing is available at Beltsville Lab http://www.ars.usda.gov/Services/docs.htm?docid=7473

This is also a stress disease. In some states you are required to burn the hive and bees and all. In some states you are required to shake the bees off into new equipment and burn the old equipment. In some states they will make you remove all the combs and bees, and they will fumigate the equipment in a large tank. Some states just require you to use Terramycin to treat them. Some states if you are treating they will let you continue but if the bee inspector finds it they make you destroy the hives. Many beekeepers treat with Terramycin (sometimes abbreviated TM) for prevention. The problem with this is that it can mask the AFB. The spores of AFB will, for all practical purposes, live forever, so any contaminated equipment will remain so unless fumigated or scorched. Boiling will not kill it. Neither TM nor Tylosin will kill the spores, only the live bacteria. AFB spores are present in *all* beehives. When a hive is under stress is the most likely time for an outbreak. Prevention is best. Try not to let hives get robbed out or run out of stores. Steal stores and bees to shore up weak hives so they don't get stressed. What you are allowed to do if you get AFB varies by state, be sure to obey the laws in your state. Personally, I have never had AFB. I have not treated with TM since 1976. If I had an outbreak I would have to decide what I would do. It may depend on how many hives are affected what I might do, but if I had a small outbreak I would probably shake the bees out into new equipment and burn the old equipment. If I had a large outbreak, I might try breaking the brood cycle and swapping out infected combs. If we as beekeepers keep killing all bees with AFB we will not breed AFB resistant bees. If we as beekeepers keep using Terramycin as a preventative we will continue to spread TM resistant AFB.

> *"It is well known that improper diet makes one susceptible to disease. Now is it not reasonable to believe that extensive feeding of sugar to bees makes them more susceptible to American Foul Brood and other bee disease? It is known that American Foul Brood is more prevalent in the north than in the south. Why? Is it not because more sugar is fed to bees in the north while here in the south the bees can gather nectar most of the year which makes feeding sugar syrup unnecessary?"—Better Queens, Jay Smith*

Parafoulbrood

This is caused by Bacillus para-alvei and possibly combinations of other microorganisms and has symptoms similar to EFB. The easiest solution is a break in brood rearing. Cage the queen or remove her and wait for them to raise one. If you put the old queen in a nuc or the old queens in a queen bank, you can reintroduce them if they fail to raise a queen.

Sacbrood

Caused by a virus usually called SBV (SacBrood Virus). Symptoms are the spotty brood patterns as other brood diseases but the larvae are in a sack with their heads raised. As in any brood disease, breaking the brood cycle may help. It usually goes away in late spring. Requeening sometimes helps also.

Breaking the Brood cycle to help with brood diseases

For all of the brood diseases this is helpful. Even for Varroa as it will skip a generation of Varroa. To do this you simply have to put the hive in a position that there is no longer any brood. Especially no open brood. If you are planning to requeen anyway, just kill the old queen and wait a week and then destroy any queen cells. Don't go three or they will have raised a new queen. Wait another two weeks and then introduce a new queen (order the appropriate amount ahead of time). If you want to raise your own, just remove the old queen (put her in a cage or put her in a nuc somewhere in case they fail to raise a new one) and let them raise a queen. By the time the new queen is laying there will be no more brood. A hairclip catcher works for a cage. The attendant bees can get in and out and the queen cannot.

Small Cell and Brood Diseases

Small cell beekeepers have reported it helping with brood diseases. Especially once the size is down below 4.9mm. We know that once a cell falls below a certain level the bees chew it out and obviously this is many more cocoons in a large cell than a small cell. (See Grout's research on this). I don't know if it helps with brood diseases or not, but my speculation (and it is merely speculation) on this is that because small cells get chewed out before a lot of cocoons build up where 5.4mm cells get filled with generation after generation of cocoons until they are down around the 4.8mm or smaller size before they get chewed out. This leaves many more places for brood pathogens to accumulate.

Neighbors

Frightened neighbors have been known to spray your hives with Raid, but usually they are too afraid to do that and just use pesticides on their flowers to get rid of bees. If they use Sevin many of your bees can die. "Courageous" neighborhood kids have been known to knock over hives in a show of bravery. Gifts of honey to neighbors and perhaps a good PR strategy help. If someone watches you open a hive with no veil it often belays their fears. But you could have the bad luck to open it on a grouchy day and get stung which only reinforces their fears. I'd wear a veil and no gloves and try not to react if you do get stung. That way they see it's not that big of a deal and the bees are not all trying to kill you.

Recent enemies

Recently new enemies have turned up.

Varroa Mites

Varroa destructor (previously called Varroa jacobsoni which is a different variety of the mite that is in Malaysia and Indonesia) are a recent invader of beehives in North America. They arrived in the USA in 1987. They are like ticks. They attach to the bees and suck the hemolymph from the adult bees and then get into cells before they are capped and reproduce there during the capped stage of the larvae development. The adult female enters the cell 1 or 2 days before it is capped, being attracted by pheromones given off by the larvae just before capping takes place. The female feeds on the larvae for a while and then starts laying an egg about every 30 hours. The first is a male (haploid) and the rest are females (diploid).

In an enlarged cell (see Chapter *Natural Cell Size* in Volume II) the female may lay up to 7 eggs and since any immature mites will not survive when the bee emerges, from one to two new female mites will probably survive. These will mate, before the bee emerges and emerge with the host bee.

Varroa

Varroa mites are large enough you can see them. They are like a freckle on a bee. They are purplish brown in color and oval shaped. If you look at one closely or with a magnifying glass you can usually see the short legs on it. To monitor Varroa infestations you need a Screened Bottom Board (SBB) and a white piece of cardboard. If you don't have a SBB then you need a sticky board. You can buy these or make one with a piece of #8 hardware cloth on a piece of sticky paper. The kind you use to line drawers will work. Put the board under it and wait 24 hours and count the mites. It's better to do this over several days and average the numbers, but if you have a few mites (0 to 20) you aren't in too bad of shape if you have a lot (50 or more) in 24 hours you need to do something or accept the losses.

Several chemical methods are available.

I think that the goal should be no treatments. But these are the common ones.

Apistan (Fluvalinate) and Checkmite (Coumaphos) are the most commonly used acaracides to kill the mites. Both build up in the wax and both cause problems for the bees and contaminate the hive. I don't use them.

Softer chemicals used to control the mites are Thymol, Oxalic acid, Formic acid and Acetic acid. The organic acids already naturally occur in the honey and so are not considered contaminates by some. Thymol is that smell in Listerine and although it occurs in Thyme honey, it doesn't occur otherwise in honey. I have used the Oxalic acid and liked it for interim control while regressing to small cell. I used a simple evaporator made of brass pipe. My concerns about all of these are their impact on the beneficial microbes in the hive.

Inert chemicals for Varroa mites

FGMO is a popular one of these. Dr. Pedro Rodriguez, DVM has been a proponent and researcher on this. His original system was cotton cords with FGMO, beeswax and honey in an emulsion. The object was to keep the FGMO on the bees for a long period of time so the mites either get groomed or they suffocate on the oil. Later a propane insect fogger was used to supplement the cords in this control system. The other up side of the FGMO fog was it apparently kills the tracheal mites also. But this could also be interpreted as a down side because you are possibly perpetuating genetics of bees who can't handle Tracheal mites.

Inert dust. The most common inert dust used is powdered sugar. The kind you buy in the grocery store. It is dusted on the bees to dislodge the mites. According to research by Nick Aliano, at the University of Nebraska, this method is more effective if you remove the bees from the hive and dust them and then return them. It is also very temperature sensitive. Too cold and the mites don't fall. Too hot and the bees die.

Physical methods.

Some methods are just hive parts or other things. Someone observed that there were less mites on hives with pollen traps and figured maybe the mites fell in the trap. The results were a screened bottom board (usually abbreviated SBB). This is a bottom board on the hive that has a hole covering most of the bottom covered with #7 or #8 hardware cloth. This allows the mites that get groomed off to fall down where they can't get back on the bees. Research shows that this eliminates 30% of the mites. I seriously doubt these numbers but I do like screened bottom boards for monitoring mites and controlling ventilation and helping with any kind of control you actually do.

What I do. I use the small cell/natural cell and I use some Screened Bottom Boards (SBB) and I used to monitor the mites with a white board under the SBB. My plan was as long as the mites stayed under control, and so far, since 2002 they have, that's all I would do. I never needed to do anything else and the mite levels dropped to where they were hard to detect. If the mites were to start going up while the supers are on I would probably remove the drone brood and maybe fog with FGMO or dust with powdered sugar. If they were still high after fall harvest, I might use Oxalic Acid vapor but I would also plan to requeen. So far I haven't needed any treatment since the bees were regressed. Just small cell has been effective for me for both kinds of mites and adequate under normal conditions.

More about Varroa

Without getting into the issue of what methods are best, I think it's significant to the success and sometimes subsequent failure of many of the methods we, as beekeepers are trying to use. I used FGMO fog only for two years and when I killed all of the mites with Oxalic acid at the end of that two years there was a total mite load of an average of about 200 mites per hive. This is a very low mite count. But some people have observed a sudden increase to thousands and thousands of mites in a short time. Part of this is, of course, all the brood emerging with more mites. But I believe the issue is also that the FGMO (and many other systems as well) manage to create a stable population of mites within the hive. In other words the mites emerging are balanced out by the mites dying. This is the object of many methods. SMR queens are queens that reduce the mites' ability to reproduce. But even if you get to a stable reproduction of mites, this does not preclude thousands of hitchhikers coming in. Using powdered sugar, small cell, FGMO or whatever that gives an edge to the bees by dislodging a proportion of the mites, or preventing the reproduction of mites and seems to work under some conditions. I believe these conditions are where there are not a significant number of mites coming into the hive from other sources.

All of these methods seem to fail sometimes when there is a sudden increase in mites in the fall.

Then there are other methods that are more brute force. In other words they kill virtually all the mites. Even these seem to fail sometimes. We have assumed it's because of resistance, and perhaps this is a contributing factor. But what if sometimes it's again because of this huge influx of mites from outside the hive? Granted having the poison in the hive over a period of time when this explosion of population occurs seems to be helpful, it still sometimes fails.

One explanation for this may be that bees robbing and drifting are causing it.

> *"The percentage of foragers originating from different colonies within the apiary ranged from 32 to 63 percent"—from a paper, published in 1991 by Walter Boylan-Pett and Roger Hoopingarner in Acta Horticulturae 288, 6th Pollination Symposium (see Jan 2010 edition of Bee Culture, 36)*

I have not had this happen on small cell... yet. Nor have I had it happen on FGMO. I have seen it happen when I was using Apistan. But others have observed it with FGMO and I have to wonder how much this affects the success of many methods from Sucracide to SMR queens, from FGMO to Small Cell. It seems like there are at least two components to success. The first is to create a stable system so that the mite population is not increasing within the hive. The second is to find a way to monitor and recover from that occasional sudden influx of mites. Conditions that cause the mites to skyrocket seem to be in the fall when the hives rob out other hives crashing from mites and bring home a lot of hitchhikers while at the same time all the mites that had been in the cells are emerging with no brood to go back into.

Tracheal Mites

Tracheal mites (Acarapis woodi) are too small to see with the naked eye. This was first called "Isle of Wight disease" as this is where it was first observed and the cause, at the time, was not known. Then when they discovered it was a mite, it was called "Acarine Disease" since it was the only known malignant mite on honeybees. Symptoms are crawling bees, bees that won't cluster in the winter and "K" wings where the two wings on each side have separated and make a shape like the letter "K". The tracheal mites have been in the US since 1984 that we know of. If you want to check for them you need a microscope. Not a really powerful one, but you still need one as they are too small to see with the naked eye. You're not looking to see the details of a cell, just a creature that is quite small.

Tracheal mites need to get into the trachea to feed and reproduce. The opening to the trachea on an insect is called a spiracle. Bees have several of these and they have a muscular system that allows the bees to totally close them if they want. Since the mites are much larger than the largest spiracle (the first Thoracic spiracle) they have to find young bees whose chitin is still soft so that they can chew out the first Thoracic spiracle enough to gain entry. Once inside, the much more spacious trachea provides the place they live and breed. Tracheal mites must do this while the bees are still 1 to 2 days old before their chitin hardens. A common control for them is a grease patty (sugar and cooking grease mixed to make a patty) because it masks the smell that the tracheal mites use to find a young bee. If they can't find young bees, they can't chew through the spiracle in old bees to get in and so they can't reproduce. Menthol is commonly used to kill the Tracheal mites. FGMO and (by some reports) Oxalic acid will also kill them. Breeding for resistance and small cell are also useful. The theory on the small cell helping is that the spiracles (the openings into the trachea) that the bees breathe through are smaller and the mites can't get in. But since they are already too small it is more likely that the smaller opening is less attractive to the mites who are looking for a hole they can enlarge enough to get in, or the chitin gets thicker the more you get from the edge and they can't chew it wide enough to gain entry. More research is needed on this subject. But basically, I'm just using small cell and they have not been a problem.

Tracheal mite resistance is not hard to breed for and may explain why small cell beekeepers aren't having any problems. If you never treat and you raise your own queens you'll end up with resistant bees. The mechanism of resistance to tracheal mites is not known. One theory is that they are more hygienic and groom off the tracheal mites before they can get in. Another is that they have either smaller spiracles or tougher spiracles that the mites can't get access through. Another could be similar to the grease patty treatment, in that the younger bees may not give off the odor that triggers the tracheal mites to seek them.

Acarapis dorsalis and Acarapis externus are mites that live on honey bees that are indistinguishable from Tracheal mites (Acarapis woodi). They are classified differently simply based on the location where they are found. Leading to the obvious question, are they the same and they are just not able to get into the trachea?

Small Hive Beetles

Another recent pest that has not been a problem where I am yet, is the Small Hive Beetle (Aethina tumida Murray). Sometimes abbreviated SHB. The larvae eat comb and honey, similar to wax moths, but are more mobile, more in groups and crawl out of the hive and into the ground to pupate. The adult beetles get the bees to feed them but the bees also like to corral them into tight corners. There is some controversy over whether these corners are bad because they give the beetles a place to hide, or good, because it gives the bees a place to corner them

The damage they do is similar to the wax moths but more extensive and they are harder to control. If you smell fermentation in the hive and find masses of crawling, spiky looking larvae in combs you may have SHB. The only chemical controls approved for use are traps made with CheckMite and ground drenches to kill the pupae, which pupate in the ground outside the hive.

While they have been identified in Nebraska, I have not had to deal with these, but I will probably go to more PermaComb in the brood nests if they become too much of a problem. Strong hives seem to be the best protection.

Some people use various traps (some homemade and some commercially available) and some people just ignore them. They seem to thrive on sandy soil and warm weather but can survive even in clay soil and nasty cold winters. How much of a problem they are, and how much effort needs to go into controlling them, seems related to those two main things: clay in the soil and cold in the winter.

Are treatments necessary?

The standard books out there on beekeeping will come across as if treating is absolutely necessary and that bees would be extinct without human intervention. Just to give you an idea, here is my complete history of treating:

1974 used Terramycin because the books scared me into thinking they would die without.

1975-1999 no treatments whatsoever but lost them all in 1998 and 1999 to Varroa

2000-2001 used Apistan for Varroa. In 2001 they all died from Varroa anyway

2002-2003 used Oxalic acid on some of them, FGMO on some, wintergreen oil on some and nothing on some of them also started regressed to small cell.

2004-present no treatments whatsoever

So the only 3 years all of my bees were treated for anything were 1974, 2000, 2001.

The only 5 years *any* of my bees were treated for *anything* would add years 2002, 2003

The 32 years that NONE of my bees were treated for ANYTHING were: 1975,1976, 1977, 1978, 1979, 1980, 1981, 1982, 1983, 1984, 1985, 1986, 1987, 1988, 1989, 1990, 1991, 1992, 1993, 1994, 1995, 1996, 1997, 1998, 1999, 2004, 2005, 2006, 2007, 2008, 2009, 2010

I look for mites (as does the inspector every year) and I look hard at deadouts to see if they died from Varroa. I see no Varroa problems anymore. I occasionally find a Varroa.

I have never treated for Nosema or purposely treated for tracheal mites (although the wintergreen and the FGMO and the Oxalic acid may have affected them)

I have bought some packages from time to time, but I was also expanding from about four hives to 200 and I was selling some small cell nucs at the same time and rearing queens.

Queen Spotting

Do You Really Need to Find Her?

I will preface this that you don't have to find the queen every time you look in the hive. In fact I have changed my methods to eliminate finding the queen as much as possible because it is so time consuming. If there is open brood then there was a queen at least a few days ago. But there are situations where you really need to find the queen. Requeening being the most likely. So here are a few tips.

Use Minimal Smoke

First, don't smoke them very much, if at all, or the queen will run and there is no telling where she will be.

Look for the Most Bees

The queen is usually on the frame of the brood chamber that has the most bees. This isn't always true, but if you start on that frame and work your way from there you will find her either on that frame or the next 90% of the time.

Calm Bees

The bees are calmer near the queen.

Larger and Longer

Of course the obvious thing is that the queen is larger, and especially that her abdomen is longer, but that isn't always easy to see when there are bees climbing all over her. Look for the larger "shoulders" The width of her back, that little bare patch on the thorax. These are all larger and often you get a peek at them under the other bees. Also the longer abdomen sticking out sometimes when you can't see the rest of her.

Don't count on her being marked

Don't count on your marked queen still being there and being marked. Remember they may have swarmed and you didn't catch it or they may have superseded and she may be gone.

Bees around the queen act differently

Look at how the bees act around the queen. Often there are several, not all, but several bees facing her. The bees around the queen act different. If you watch them every time you find a queen you'll start noticing how they act, and how they move different around her.

The Queen Moves Differently

Other bees are either moving quickly or just hanging and not moving. The workers move like they're listening to Aerosmith. The queen moves like she's listening to Schubert or Brahms. She moves slowly and gracefully. It's like she's waltzing and the workers are doing the bossanova. Next time you spot the queen notice how the bees in general move, how the bees around her move and how she moves.

Different Coloring

Usually the queen is slightly different color. I have not found this helpful because she's also usually close enough in color that she's still hard to spot by this.

Believe there is a Queen

Also, mental attitude makes a difference when trying to find anything from your car keys to hunting deer to finding a queen. As long as you are doing cursory looks thinking it won't be there you won't find it. You have to believe that the keys, or the deer or the queen *is* there. That you are looking right at it and you just have to see it. And then suddenly you do. You have to convince yourself that it is there and convince yourself that you will find it. I don't know how to explain it well enough, but you have to learn to think like that.

Practice

Of course the best solution to learning to find a queen is an observation hive. You can find one every morning when you get up, every evening when you get home, and every night before you go to bed and not disrupt them at all. It still doesn't give you the practice at finding the right frame on the first try or two, but does help you with spotting her. Having the queen marked in the observation hive is nice for showing the queen to visitors, but *not* having her marked works better for practicing finding the queen. Even if you buy all your queens marked you will often be finding an unmarked supersedure queen.

Can you find her?

Here she is.

How about this one?

Does this help?

Fallacies

I'm sure some people believe these and will dis-agree, but here are some ideas that I consider myths of beekeeping:

Myth: Drones are bad.

Drones, of course are normal. A normal healthy hive will have a population in the spring of somewhere around 10-20% drones. The argument for almost a century or more (really just a selling point for founda-tion) was that drones eat honey, use energy and don't provide anything to the hive, therefore controlling the drone comb and therefore the number of drones will make a hive more productive. All the research I've heard of says the opposite is true. If you try to limit the number of drones your production will decrease. Bees have an instinctive need to make a certain number and fighting that is a waste of effort. Other research I've seen says that you will end up with the same number of drones no matter what you, the beekeeper do anyway.

Myth: Drone comb is bad.

This, of course, goes with the first one. The way a beekeeper attempts to control drones is by having less drone comb. But controlling drone comb is exactly the reason you end up with drone comb in your supers and then end up needing an excluder. The bees want a consolidated brood nest, but the lack of drone is more worrisome to them, so if you don't let them do it in the brood nest, they will raise a patch of drones anywhere they can get some drone comb. If you want the bees to stop building drone comb, stop taking it away from them. If you want the queen to not try to lay in the supers, let them have enough drone comb in the brood nest.

Myth: Queen Cells are bad

...and the beekeeper should destroy queen cells if they find them.

It seems like most of the books I've read convince beginning beekeepers that queen cells should always be destroyed. The bees are either going to swarm, and you want to stop them, or they are trying to replace that precious store-bought queen with a queen of unknown lineage mated with those awful feral drones. Most of the time when you destroy queen cells the bees swarm anyway, or they already swarmed before you destroyed them, and they not only swarm, but also end up queen-less. I see swarm cells as free queens of the highest quality. I put each frame that has queen cells on it, in its own nuc. Usually I try to leave one with the original hive and the old queen in a nuc. That way I've made a bunch of small splits and left the hive thinking it's swarmed already. With supersedure cells, I leave them because the bees apparently have found the queen wanting and I trust the bees. Destroying a supersedure cell is also likely to leave them queenless. The queen is probably about to fail, or she's already failed or died and you just removed their only hope of a queen.

Myth: Home grown queens are bad

...and beekeepers should buy queens because mating with the local bees is bad.

Of course this one goes with the above reasons given for why supersedures are bad. I think mating with the local bees is the preferred method. You get bees that are surviving in your area. I do know a lot of people who buy queens all the time because of this fallacy. The supersedure rate has grown over the years to the point that a typical introduced queen is almost instantly superseded. If that's true (and some of the experts tell me it is) then you'll have a home grown queen anyway, so why waste your money? There is a lot of research on how much better the quality of a queen is if you let her continue to lay from when she starts instead of banking her right after she starts laying. When you buy a commercial queen, you get one that was banked right after she started to lay. I have serious doubts that you can buy a better queen than you can raise yourself, especially if you have clean wax; and most especially if you've been collecting swarms from bees that live in your climate.

Myth: Feral Bees are bad.

...feral bees are unproductive, swarmy and of bad disposition.

I've heard this often repeated—this or other disreputable things. Feral bees probably were at one time but lately have not been, bred for disposition. I've removed and caught many. Some are mean. Some are quite nice. Some are nervous, but not mean. Some are calm. These traits I have found easy to find in feral bees and easy to breed for. Just keep the good ones and requeen the bad ones. From my experience they are often more productive because they are more attuned to your climate and build up at the appropriate time to make a good crop. As far as "swarmy" I think all bees are swarmy. It's how they reproduce. I have not had any problems controlling swarming in any kind of bees.

Myth: *Feral swarms are disease ridden*

... and should either be left, killed or treated immediately by the beekeeper for every known disease.

I don't understand the concept. A healthy productive hive throws swarms. So the logical conclusion would be that they are healthy and productive.

Myth: *Feeding can't hurt anything.*

I hear this one a lot. But I think feeding *can* hurt a lot. Feeding is one of the leading causes of problems. It attracts pests like ants, it sets off robbing, it often drowns a large number of bees, and worst, it often results in a nectar bound brood nest and swarming. If the hive is light in the fall, the beekeeper should feed. If the bees are starving, feed. If you're installing a new package or a swarm, feed until they get some capped stores. But once they have a little stored and there's a flow, let them do what bees do—gather nectar. A good rule of thumb is that they should have at least some capped comb and a flow before you stop feeding.

Myth: *Adding supers will prevent swarming.*

This is a common myth in beekeeping. It works after the reproductive swarm season is over, but the prime swarm season has little to do with supers. It has everything to do with the bees' plan to reproduce. If you want to head off a swarm the crux of the matter is you have to keep the brood nest open. Part of that plan is to put on supers before they backfill the brood nest but that alone cannot be relied on to stop them from swarming.

Myth: Destroying queen cells will prevent swarming.

In my experience this does not work. They will swarm anyway and end up queenless.

Myth: Swarm cells are always on the bottom.

The other part of this, I guess is that supersedure cells are always in the middle. This may be a good generality, but you need to look at the entire context of the situation. I would assume that queen cells on the bottom were swarm cells if the hive is building up quickly and is either very strong or very crowded. On the other hand if they are not strong or crowded and building, then I would assume they are not swarm cells. If the cells are more in the middle and conditions otherwise would cause me to expect swarm cells, then I would tend to view these as swarm cells. If the hive were not building and not crowded I would assume they are supersedure cells or emergency cells. Also swarm cells tend to be more numerous.

Myth: Clipping the queen will prevent swarming.

In my experience they will still swarm. It may buy you some time if you're paying attention (like the hives are in your back yard and you check every day for swarms). They will attempt to swarm and the clipped queen won't be able to fly. They will go back and then they will leave with the first virgin swarm queen to emerge. Counting on clipping to stop them from swarming will end in failure.

Myth: 2 Feet or 2 miles.

...you have to move hives two feet or two miles or you will lose a lot of bees.

I hear this one a lot. Anytime you move bees there will be some chaos for at least one day, but I move bees all the time fifty, a hundred yards or more. The trick is to put a branch in front of the entrance to trigger reorientation. If you do this it works well. If you don't do this most of the field bees will go back to the old location. That and accept that there will be some confusion for a while, so don't move them if you don't have a reason.

Myth: You have to extract.

...or that it's somehow cruel to the bees to not extract.

The beginner beekeepers all seem to think they have to buy an extractor. It's not their fault. It's what the books all say, right? You don't. I had bees for 26 years without one. You can make cut comb or crush and strain with little investment and no more work than extracting.

Myth: 16 pounds of honey = 1 pound of wax.

This is an old one that is still sold to beekeepers at various numbers. I know of no study to support it. And it's irrelevant. What is relevant is how productive a hive is with and without drawn comb. There is no doubt they will make more honey with drawn comb. But it would take a lot of hives before it would be worth buying an extractor. This concept is also used to sell foundation. In my experience the bees will draw comb faster without foundation than with it and the faster they have somewhere to store the nectar the more honey they make.

Myth: You can't raise honey and bees.

...in other words, make splits and get production.

It's all in the timing. If you do the split right before the flow and let all the field bees drift back to the original hive you can actually get more honey and more bees.

Myth: Two queens can't coexist in the same hive.

People purposely set up two queen hives all the time. But if you look carefully you'll often find two queens naturally in a hive. Usually a mother daughter, where the supersedure queen is laying and the old queen is laying right beside her.

Myth: Queens will never lay double eggs.

...in other words, all multiple eggs are a sign of a laying worker.

I've often seen double eggs from a queen. Rarely I've seen triples. I've seldom seen more than triples. Laying workers will lay from two to dozens in one cell. I look for more than two and eggs on the sides of the cells and not in the bottom. Also eggs on pollen. These I consider signs of laying workers.

Myth: *If there is no brood there is no queen.*

There are many reasons you might find a hive with no brood even though there is a queen. First, in my climate at least, from October to April there may or may not be brood because they stop in October and then raise little batches of brood with broodless periods in between. Second, some frugal bees will shut down brood rearing in a dearth. Third, a hive that has lost a queen and raised an emergency queen often is brood-less because by the time the new queen has emerged, hardened, mated and started to lay 25 or more days have passed and *all* the brood has emerged. Fourth a hive can swarm and the new queen isn't laying yet. She won't be laying for probably at least three weeks after the hive swarmed. Many a beginner (or even a veteran) beekeeper has found a hive in this state, ordered a queen, introduced her and had her killed, ordered another queen, introduced her and had her killed and finally noticed there were eggs. Unmarked virgin queens are very hard to find even by the most experienced beekeeper. A frame of eggs and brood would have been a better insurance policy. That way *if* the hive is queenless they can raise one, and if they aren't it won't hurt anything and you'll know the answer to the question. See the section *Panacea* in the Chapter *BLUF.*

Myth: *Bees only like to work up.*

...in other words they expand the hive and the brood only in an upward direction and not downwards.

If you install a package in a stack of five boxes, as I have done on occasion, you can easily disprove this. But then if you think about a swarm in a tree you already know this isn't true. The bees cluster at the top of whatever space there is and build comb down until they fill the void or reach a size they are satisfied with.

Bees start at the top of whatever space they have and work down. In a tree there is no other choice as there is no way to work up. Once a hive is established they move towards any space they can fill. So in the case of a tree if they have reached the bottom the brood nest will work its way into whatever space is available when it expands and then contract back when the season is over. In the case of a hive, however, beekeepers keep adding and removing boxes. We add them to the top because it's convenient to add them there and convenient to check on them there. The bees don't care. They work into where there is space available.

Myth: A laying worker hive has one pseudo queen

.. and you are trying to get rid of her to fix the problem.

A laying worker hive has many laying workers. The only way to fix the problem is get them so disrupted they will accept a queen or give them enough pheromones from open worker brood to suppress the laying workers enough to get them to accept a queen. In other words, give them a frame of open brood every week until they start to rear a queen. Then you can either let them finish or introduce a queen.

Myth: Shaking out a laying worker hive works

...because the laying worker gets left behind because she doesn't know her way home.

I have not found this to be true and the research I've read says it's not true. There are many laying workers and they will have no trouble finding their way back. Shaking out a hive only works sometimes because you have disheartened them enough that in the chaos they will sometimes accept a queen.

Myth: Bees need a landing board.

Obviously they don't have one in most natural situations, so this is not a rational statement. I not only don't think they need one, I think they just help mice and skunks and do no favors to the bees.

Myth: Bees need a lot of ventilation.

Bees do need ventilation. But what they need is the right amount of ventilation. Of course in the winter, too much ventilation means too much heat loss. But even in the summer the bees are cooling the hive by evaporation, so on a hot day the inside of the hive may be cooler than the outside air. So too much ventilation could result in the bees being unable to maintain a cooler temperature inside. When wax heats up past the normal operating temperatures of a hive (> 93° F or 34° C) it gets very weak and combs can collapse.

Myth: Bees need beekeepers.

Actually bees need beekeepers like "fish need bicycles." Depending on your view of the world, bees have been surviving for millions of years on their own or at least since the creation. Its true beekeepers have spread them all over the world, but bees would have gotten there anyway eventually. How did African bees recently get to Florida? They were hitchhikers.

Myth: *You have to requeen yearly.*

I know many beekeepers who only requeen if they see a problem. Usually before you see a problem the bees have already superseded the problem queen. If they have, you have perpetuated genetics that know how to do this. If you have clean wax (no chemicals in the hive) your queens usually last about three years. If you don't have clean wax, your queens usually only last a few months. Either way, how does requeening yearly help? The most common claim is that a first year queen won't swarm, which is easily disproven by feeding a package incessantly, or that a second year queen is bound to swarm, which is easily disproven by the fact that most of my queens are three years old.

Myth: *A marginal colony should always be requeened.*

I've seen a lot of struggling colonies take off and make a good crop. They are often struggling because the population dwindled to the point that there weren't enough workers to forage and care for brood. Quite often a frame of emerging brood will snap them right out of this. On the other hand *some* colonies do just languish when they should have caught up. These I would requeen.

Myth: *You need to feed pollen substitute*

...to packages and to bees in the spring and fall.

I have never had luck getting bees to even take pollen once fresh pollen is available. I see no reason to feed a package pollen substitute when it is vastly inferior nutrition to real pollen that is readily available that time of year. Feeding real pollen early in the spring sometimes seems to be an effective way to stimulate buildup. Sometimes it seems to make no difference.

Myth: You should feed syrup in the winter.

I suppose your climate is directly related to this, but you can't get bees to take syrup in the winter here in Nebraska and if you could, I'm not sure it would be good for them to have all that humidity to deal with. Dry sugar they can take no matter how cold it is, but syrup they can only take if the syrup is above 50º F. Not a likely occurrence here even if the daytime temps got up to that, the syrup would have a time delay making it up to that temperature.

Myth: You can't mix plastic and wax.

This is not so much a myth as an over simplification. Putting undrawn plastic in with undrawn wax is like putting a piece of cherry pie and a bowl of broccoli in front of your kids at the same time. If you want them to eat the broccoli, you should wait to put out the cherry pie.

If you mix wax and plastic foundation, the bees will jump on the wax and ignore the plastic. If you put in all plastic they will use it when they need comb.

There is no great impending disaster if you mix them. They just have their preferences and if you want them to follow *your* preferences you should limit their choices.

Once it is drawn comb or comb that is being used, you can mix it freely with everything with no problems.

Myth: Dead bees headfirst in cells have starved.

This is a commonly held belief. All dead hives over winter will have many bees with their heads in cells. That's how they cluster tightly for warmth. I would read more into whether or not they are on contact with stores.

Realistic Expectations

I think it's important in every aspect of beekeeping to have realistic expectations. Not to say that those may not be exceeded at times, but also at times they will not be met as both failure and success are dependent on many related variables.

As examples, let's consider some of the variable outcomes.

Honey Crop

Typically people tell beginner beekeepers not to expect a honey crop the first year. This is an attempt to set realistic expectations. However a good package with a good queen in a good year (appropriate amounts of well timed rainfall and flying weather) may far exceed this or may not even get well established. But generally it's a realistic expectation for the beekeeper that they should get established enough to get through the winter and maybe make a little honey.

Plastic Foundation

People buy plastic foundation (and other plastic beekeeping equipment such as Honey Super Cell fully drawn comb) and sometimes are very disappointed. The bees typically will hesitate to draw the plastic (or use the Honey Super Cell) and this sets them back a bit. Sometimes the bees will draw a comb between two plastic foundations in order to avoid using it. Sometimes they will build "fins" out from the face of the foundation. None of these are unusual, but they also often draw it pretty well. How well they do depends on a combination of genetics and nectar flow. Many people seeing the hesitation decide never to use plastic again. But actually once the bees use it, comb on plastic foundation or even fully drawn plastic comb is used just like any other comb. The delay at first seems like a big setback, and for a package, perhaps it is, but once you get past it there is no problem getting it used after that.

Wax Foundation

People use wax foundation and often it gets hot and buckles, or the bees chew it all up or the bees don't want to draw it and they drawn fins or combs between. They do this less with than with plastic, but still sometimes they do. The buckled foundation often gets comb built on it and the comb is a mess. Many people after an experience like this say they will never use wax foundation again. But really that's just how the circumstances went. If you put it in on a good flow the bees would not have chewed it and it would have been drawn before it buckled. My point is that people often have unrealistic expectations and when those are not realized, they are disappointed in the method when it was other circumstances that led to the problems.

Foundationless

Some people use foundationless frames. Many have perfect luck with it but some will have bees that just don't get the concept and build some crossways comb. Since this happens just as often in plastic foundation, and wax foundation that has collapsed or fallen out etc. it would not seem that significant to me, but if the only experience you have is with the foundationless, you may assume that other methods don't have these problems. But they do. Again, genetics and timing of the flow have a lot to do with success or failure.

The most important concept to grasp with any natural comb hive is that because bees build parallel combs, one good comb leads to another in the same way that one bad comb leads to another. You cannot afford to not be paying attention to how they start off. The most common cause of a mess of comb is leaving the queen cage in as they always start the first comb from that and then the mess begins. I can't believe how many people want to "play it safe" and hang the queen cage. They obviously can't grasp that it is almost a guarantee of failure to get the first comb started right, which without intervention is guaranteed to mean every comb in the hive will be messed up. Once you have a mess the most important thing is to make sure the *last* comb is straight as this is always the guide for the *next* comb. You can't take a "hopeful" view that the bees will get back on track. They will not. You have to put them back on track.

This has nothing to do with wires or no wires. Nothing to do with frames or no frames. It has to do with the last comb being straight.

Losses

New beekeepers often assume that every hive should live forever and every hive should make it through the winter. Some winters, they do. But most winters kill off at least a few of the hives. Obviously the more hives you have the more this happens. I went years without losing a hive, but I only had a few and I always combined any that were borderline on strength and those were the days before Tracheal mites, Varroa mites, Nosema ceranae, small hive beetles, and a host of viruses we now have. Now I have around two hundred hives and try to overwinter a lot of nucs, of marginal strength and there are those many new diseases and pests to stress them out. No winter losses are an unrealistic expectation. But high winter losses are a sign that you must be doing something wrong or the weather did something very quirky.

I always try to figure out the cause of winter losses. Often it is starvation from getting stuck on brood. Sometimes with nucs or small clusters it's a hard cold snap (-10 to -30 F) and the cluster just wasn't big enough to keep warm. I always look for dead Varroa. Finding thousands of dead Varroa in the dead bees is a good indication that the Varroa were the primary cause of their death. A lack of such evidence is probably good evidence that it was something else.

Again, the point is that sometimes wintering exceeds or falls below even realistic expectations. But it's helpful to start with realistic expectations and work from there. Realistic expectations from healthy hives as far as losses are probably in the 10% range with some years worse and some years better.

Splits

One of the common questions I hear from new beekeepers is "how many splits can I make?" Of course the answer to this is probably the most variable of any except, perhaps, "how much honey will my hive make?" The difference between a good year and a bad year in beekeeping varies far more than 10 fold. I've had years where I got 200 pounds of honey from every hive and years where I harvested nothing and fed 60 pounds of sugar (between spring and fall) to every hive. Splits are similar. Some hives can't be split at all. Some can be split five times in a year. Most can only take one split and still make a decent crop of honey and be well stocked for winter.

The point of all of this is that results in beekeeping vary dramatically based on what is happening around the bees as well as things like the time of year, the way they are cared for and so on. It's very difficult to predict what the outcomes will actually be, so there is no point in having too high or low of expectations. Take things as they come and adjust. Be prepared for both exceptional success and failure and adjust as you go.

Harvest

Beginners are often convinced they must have an extractor. There are many other options that make more sense. One would be comb honey.

Comb Honey

Normally I'm not shy about saying things my own way, but Richard Taylor said this so well, I will not even attempt to do better. For more of his wisdom check out his books including *The How to do it book of beekeeping*, *The Joy of Beekeeping* and *The Comb Honey Book*.

Richard Taylor on comb honey and extractors:

> "...time after time I have seen novice beekeepers, as soon as they had built their apiaries up to a half dozen or so hives, begin to look around for an extractor. It is as if one were to establish a small garden by the kitchen door, and then at once begin looking for a tractor to till it with. Unless then, you have, or plan eventually to have, perhaps fifty or more colonies of bees, you should try to resist looking in bee catalogs at the extractors and other enchanting and tempting tools that are offered and instead look with renewed fondness at your little pocket knife, so symbolic of the simplicity that is the mark of every truly good life."

Expense of making wax

Richard Taylor on the expense of making wax:

> *"The opinion of experts once was that the production of beeswax in a colony required great quantities of nectar which, since it was turned into wax, would never be turned into honey. Until quite recently it was thought that bees could store seven pounds of honey for every pound of beeswax that they needed to manufacture for the construction of their combs—a figure which seems never to have been given any scientific basis, and which is in any case quite certainly wrong."*

From *Beeswax Production, Harvesting, Processing and Products*, Coggshall and Morse pg 35

> *"Their degree of efficiency in wax production, that is how many pounds of honey or sugar syrup are required to produce one pound of wax, is not clear. It is difficult to demonstrate this experimentally because so many variables exist. The experiment most frequently cited is that by Whitcomb (1946). He fed four colonies a thin, dark, strong honey that he called unmarketable. The only fault that might be found with the test was*

*that the bees had free flight, which
was probably necessary so they
could void fecal matter; it was
stated that no honey flow was in
progress. The production of a pound
of beeswax required a mean of 8.4
pounds of honey (range 6.66 to
8.80). Whitcomb found a tendency
for wax production to become more
efficient as time progressed. This
also emphasizes that a project
intended to determine the ratio of
sugar to wax, or one designed to
produce wax from a cheap source of
sugar, requires time for wax glands
to develop and perhaps for bees to
fall into the routine of both wax
secretion and comb production."*

The problem with most of the estimates on what
it takes to make a pound of wax is they don't take into
account how much honey that pound of wax will support

From *Beeswax Production, Harvesting, Processing
and Products*, Coggshall and Morse pg 41

*"A pound of beeswax, when made
into comb, will hold 22 pounds of
honey. In an unsupported comb the
stress on the topmost cells is the
greatest; a comb one foot (30 cm.)
deep supports 1,320 times its own
weight in honey."*

Crush and Strain

I kept bees for 26 years without an extractor. I made cut comb honey and I did crush and strain to get liquid honey. When I finally did buy one I got a motorized radial 9/18 (holds 9 deeps or 18 mediums).

The method I arrived at to crush and strain is a double bucket strainer. I use these even when I'm extracting because they hold so much honey and it's the only way I can keep up with straining as I go.

Making the top bucket for the double bucket strainer. Drill the holes. If you make the holes small enough you can just use the bottom of the bucket for the strainer with no other strainer or screen. You can skim the wax off the top and leave whatever settles on the bottom. Cut the middle out of a lid (leaving an inch rim for the top bucket to rest on).

Using the double bucket strainer to strain honey.

Extracting

Extracting is a process where the caps are cut off of the combs and they are spun in a centrifuge called an extractor.

Cutting cappings off.

Cutting low spots.

Loading the extractor

Extracting.

Removing bees for harvesting

This is always a topic rife with disagreement. A lot of this is due to personal experience. Timing of these methods changes the outcomes tremendously.

Abandonment

C.C. Miller's favorite method is usually called "abandonment". This is where you pull each box off the hive and set it on its end so the top and bottom are exposed. This is best done at the end of the flow but not during a dearth and just after sunset but before dark. The bees tend to wander back to the hive and you can take the supers. If there is brood in them, they will not leave. If there is a dearth you will set off a robbing frenzy. If you do it in the middle of the afternoon this will be harder to deal with. This requires handling the boxes twice. Once to take them off and once to load them up. (I'm not counting the rest of the process)

Brushing and/or shaking

Some people just pull each frame, shake or brush off the bees and put the frame in a different box with a cover. This puts many bees in the air and is a bit intimidating and is tedious. You move every box a frame at a time and then you load the boxes a box at a time.

Bee Escapes

There are several kinds and the results may vary based on the kind. I never had any luck with the Porter escapes that go in the hole on the inner cover. But I have liked the triangular ones from Brushy Mt. Usually the supers are removed, the escape is put on (it's one way so be sure it's the right way, letting the bees out, but not in) and you wait a day or two for the bees to leave. Again, they will not leave if there is brood in the supers. I prefer to put one of these on a bottom board (with the escape down) and stack supers up about as high as I can reach them and then put one on top (with the escape up) and leave them overnight. If you live in SHB territory, I would not leave them longer. The biggest disadvantage is you have to handle every box three times if you put it on the hive (once to get them off, then put on the escape, then stack them back on the hive, then load them up) and twice if you put it on its own bottom board (once to stack them on the bottom board and once to load them up).

Blowing

The concept is to just blow all the bees off the combs. Some people use a leaf blower and some buy a bee blower. One argument against is that anything strong enough to blow the bees off will rip many of them in half. I've never used it so I can't say.

Butyric

I listed this separate from Bee Quick although they have some things in common, I don't consider them even in the same ballpark. Both are bee repellents that are used to drive the bees from the supers. Bee Go and Honey Robber are Butyric which is not a food safe chemical and smells like vomit. Honey robber smells like cherry flavored vomit. The chemical is put on a fume board, which is put on top of the hive. The bees are driven down and the supers are pulled off and loaded. They are only handled once. I have smelled it. I have never used it.

Fischer Bee Quick

Jim Fischer doesn't want to give away his trade secrets so he won't say what's in this. But is smells like benzaldehyde to me. Benzaldehyde is the smell of Maraschino cherries or almond extract. After making benzaldehyde in my organic chem. class, I've never been able to eat a Maraschino cherry again. It's also the main ingredient in artificial almond flavoring. But Jim Fischer assures us this is nothing but food grade essential oils. It certainly smells better and, by all accounts, is much safer than butyric. Otherwise it works on the same principle. You put it on a fume board on top and drive the bees down. The supers only have to be handled once to load them. I have smelled it and it smells fine, but I have never used it.

Frequently Asked Questions

As a moderator and participant on various bee forums, I hear these questions often, so I thought I would address some of them here.

Can queens sting?

I've been handling queens off and on since 1974. Since I started rearing queens in 2004 I've been handling hundreds of them a year. I've never been stung by a queen. I have seen them go through the motions though.

Jay Smith, a beekeeper who reared thousands of queens a year for decades, said he was only stung by one once and he said she stung him right where he had squished a queen earlier and he though she thought it was a queen.

Can they? Yes. Will they? Extremely doubtful. The few people who I've met who say they've been stung by a queen say it didn't hurt as bad as a worker.

What if my queen flies off?

This often comes with other questions like, she flew off what are the odds she came back. First let's look at what to do. If the queen flies, the first thing you do is stand still. She will orient on you and probably find her way back. The second thing to do is encourage the bees to guide her back with Nasonov pheromone. To do this, take a frame out that is covered in bees and shake them back into the hive. This will cause them to start fanning Nasonov. Third, if you don't see the queen fly back in (be watching and you may) then wait ten minutes with the cover of the hive off so she can smell the Nasonov. If you do these three things the odds are very good she will find her way back.

If you didn't do those things, there is probably a little better than 50/50 chance she will find her way back anyway.

How do you avoid her flying? Keep a close eye when you're popping the cork. Queens are fast. If you put the cage on top of the pile of bees you just dumped in the hive, she and that is down in the hive and you are bent over the top of the hive, then she is less likely to fly up.

Dead bees in front of the hive?

With the queen laying 1,000 to 3,000 eggs a day and bees living about six weeks, there are *always* some dead bees in front of the hive. Often you don't see them because they are in the weeds or grass. A *lot* of dead bees (piles of them) might be cause for concern because it may be a sign of pesticide poisoning or some other problem. But some are normal.

Frame spacing in supers and brood nests?

This question seems to come up a lot. The question is usually something like "should I put 9 or 10 frames in my supers?" or "should I put 9 or 10 frames in my brood boxes?"

My answer for the brood boxes is that I put 11 in. At least in a ten frame box. I shave the ends down in order to do this and I do it because it's the spacing the bees use if you let them. But 10 will do. They should be tightly together in the center, and not spaced out evenly. They are already further apart than the bees would prefer and spacing them any further usually results in burr comb or even an extra comb in between the frames. The theory of doing 9 in the brood box is that there will be more cluster space, less swarming and less rolling of bees. The reality, in my experience, is that it requires more bees to keep the brood warm, the sur- face of the combs is more irregular and this causes more rolling of the bees when removing frames. This irregularity is due to the fact that honey storage comb can vary in thickness but brood comb is always the same thickness. The results are that where they have honey and you have 9 frames, they have extra room to fill and they fill it with honey. If they have brood then they are not as fat as when they have honey in them. I tried 9 frames in the brood nest and was not impressed. I now have eight frame boxes and I have 9 frames in them (which requires shaving the end bars down). At 11 in a ten-frame box you get very flat consistent comb and you get smaller cell size more easily.

My answer for the supers is that *once they are drawn* you can put 9 or even 8 in the ten frame supers with good effect as the combs will just be thicker. But when it's bare foundation, the bees will often mess up the comb if you space it more than ten. Ten frames of bare foundation should always be tightly together in the middle of either a super or a brood box in order to prevent the bees from attempting to build a comb between the foundations instead of on them. With eight frame boxes you can do seven drawn combs or even six.

A related issue is messed up combs.

Why do the bees mess up the combs?

Some of this is genetics. Some bees build straight parallel combs no matter what you do. Some will burr things up every which way no matter what you do. But there are things you can do to stack the deck.

Some of it is giving them the freedom to mess it up. Push all the frames tightly together. Those spacers on the frames are there for a reason. Use them. Do not space the frames evenly in the box. When you have undrawn foundation, do *not* space less frames in a box. Bees, if they don't like your foundation (and they never do really) and if you give them the room (by spacing the combs more than $1^3/_8$" apart) will try to build a comb between two frames rather than build it on your foundation. So pushing it together makes the space between the foundations small enough to discourage this, as it's not enough room for a brood comb.

Some of it is that they don't like you deciding their cell sizes. They will build their own comb with much more enthusiasm than they will build foundation. So they try to avoid building on the foundation. One solution is to stop using foundation and go foundation- less. Another is to get foundation that is closer to what they wanted to build. 5.4mm standard foundation is much larger than typical natural worker brood comb. 4.9mm is closer.

They usually don't like plastic much. The solution to getting them to draw it is to give it to them when they need to draw comb. Don't give them wax founda- tion mixed with plastic foundation or they will ignore the plastic and draw the wax. Buy the wax coated plastic so they will accept it better. Spray some syrup on it or syrup with essential oils like Honey Bee Healthy, to cover the smell of the plastic. Once they've licked it clean they tend to accept it better.

Sometimes they will still mess it up.

How do I clean up used equipment?

Used equipment has been a controversial subject for more than a century. AFB (American Foulbrood) is still an issue but used to be an even bigger issue. The only real concern about used equipment is AFB. AFB spores live virtually forever (longer than us anyway) and infected equipment is probably one of the contributing factors to getting AFB. Many people with AFB just burn the equipment. Some scorch it. Some boil it in lye. Some "fry" it in paraffin and gum rosin.

So the issue usually is that you have at your disposal (either free or cheap) used equipment. Cleaning up from mice isn't too complicated. Just leave it out in the rain until it smells ok. Cleaning up from wax moths is just cutting out the webs (which are hard for the bees to remove) and scraping off the cocoons. If combs are dry and brittle, let the bees fix them, they will be fine. If they are dusty, the bees will clean them up. The real risk is AFB. If you have old brood comb, I would look for scale in the bottom of the cells which would indicate AFB. If there is scale, you'll have to take the threat of AFB pretty serious. Some would just burn at that point. So, assuming you find no scale then what do you do? I can't tell you what to do as it is always a risk and if you get AFB I don't want you blaming me. But I'll tell you what I do. I've always gotten mine from sources I believed to be honest, usually very cheap or free and just used the equipment with nothing done to it. I've never gotten AFB in my hives.

Now that I'm dipping my equipment, I would dip any used equipment, since I have the wherewithal.

How do I prepare the hive for winter?

More detail on this in Volume 2 *Wintering Bees* Chapter

The problem with answering this question is that it will depend on your location. There is a big difference in the issues faced by a beekeeper in South Georgia or Southern California, compared to one in Northern Minnesota or Anchorage Alaska.

So I can only give a generalization and call on my own experience in the middle of the country. I'm in Southeast Nebraska and used to be in Western Nebraska and the front range of the Rockies. So this advice is pretty useful in that range of climates.

Reduce the space. There is no reason to have extra empty space in a hive in the winter in the North. Any box that is empty combs or foundation I would pull off for the winter.

Block the mice. Mice can devastate a hive. Make sure if you have bottom entrances that you have mouse guards on. A piece of #4 hardware cloth works well for this.

Remove excluders. If you use excluders they need to be off before winter sets in. A queen can get stuck on the other side of the excluder and die in cold weather.

Make sure you have some kind of top entrance. I like all top entrances and no bottom entrances, but regardless you need at least a small one for release of moist air so you don't get condensation on the lid and so the bees can get out when the snow is deep or there are too many dead bees on the bottom board. Commonly people ask if the heat won't all escape. Heat is seldom the issue it's the condensation dripping on the bees that usually kills bees in winter.

Make sure they have enough stores. In my part of the country with Italian bees you need the hive to weigh about 150 pounds for good insurance for the winter. They probably will get by on 100 pounds, but they could also burn through that in the spring rearing brood and come up short. Any less than 100 pounds would worry me a lot. The time to feed is when the weather is still warm as they won't take syrup after it gets cold. Once you hit the target weight there is no need to feed anymore. Usually a 150 pound hive around here is two deep ten frame boxes, or three medium ten frame boxes or four medium eight frame boxes, mostly full of honey.

I have only wrapped once and was not favorably impressed, but if it's the norm for beekeepers where you live you might want to consider it. The normal wrap is 15# roofing felt as it provides some heat gain on sunny days. I found this sealed in too much moisture. Other wraps are wax impregnated cardboard and one style leaves an airspace around the hive. This seems like a wiser choice for the moisture issue. If I were trying again I'd either use the cardboard with an air-space or tack some one bys on the corners first and then use the felt with that airspace.

Avoid the temptation to think that heating a nor-mal strong hive is helpful. It's really not. Thick insula-tion is not either. With insulation they won't get warm on a sunny day and do a cleansing flight. Don't move them indoors, they need to fly. Don't pile bales of straw around as it will just attract the mice. A wind break is nice if you can provide one. If you're using straw bales for this, build that wall a ways from the hives.

How far do foragers fly?

According to Brother Adam he had bees he knew flew five miles or more to gather heather nectar. Ac-cording to Huber, he marked workers, took them different distances and released them and looked for them to turn back up at the hive. He said they always found their way back when they were $1^1/_2$ miles from the hive, but past that they didn't. He also says, and it makes sense, that it would depend on the foraging available. It also seems to vary by bee size. Brother Adam says his native Apis Mellifera mellifera, which were smaller, flew the five miles to get the Heather, but the Italians he replaced them with, which were larger, would not. Dee Lusby says her small cell bees, after regression, came back with totally different pollens than before and that based on the blooms and the spread of flora that depend on pollination she's confident that the small cell bees forage much further than the large cell bees. This would be consistent with Brother Adam's observations.

How far do drones fly to mate?

I don't think anyone really knows. They fly to DCAs (Drone Congregation Areas) and there are certain topographical clues they look for as well as pheromone trails in order to find one. DCAs are usually at a place where a tree row meets a tree row. The research seems to show that drones fly to the nearest DCA. The location, being dependant on the terrain and the amount of other hives nearby, the distance is hard to predict. Most of the scientists, however, say they fly, on the average, a shorter distance than the queens.

How far do queens fly to mate?

As with many questions with bees, it's such a variable thing to start with; it's hard to say. According to Jay Smith, who tried an island for his mating yard, and he says the queens flew at least as far as two miles.

Some estimates I've seen are as much as four or five miles. But I've also heard beekeepers who say they've seen matings (as evidenced by drone comets and the queen returning to the mating nuc) that occurred right in the beeyard.

How many hives can I have on one acre?

The problem with this question is it assumes the bees will stay on the 1 acre. They will forage the surrounding 8,000 acres.

How many hives can I have in one place?

Another common question about beekeeping is "how many hives can I put in one place?" With awesome forage (like in the middle of 8,000 acres of sweet clover), and good weather, it may be close to impossible to put too many in one place. With poor forage and drought, it may be that only a few hives is too many. A typical number that is thrown about is 20. This is a nice round number that is applicable as a generality, but to be realistic it will depend on many things and many of those things vary from year to year.

How many hives to start with?

The standard answer for a beginner is two. I'll say two to four. Less than two and you don't have resources to resolve typical beekeeping issues like queenlessness, suspected queenlessness, laying workers etc. More than four is a bit much for a beginning beekeeper to keep up with.

Planting for bees

Beekeepers always seem to want to know what to plant for their bees. Just make sure you understand that your bees will not just work the flowers on your land. They will be foraging a 2 mile radius which is 8,000 acres. It's difficult, unless you own that 8,000 acres, to plant enough to make a crop. But it's not hard to plant things that will fill out the year for the bees. The times of need in the hives is early (February to April), late (September to the killing frost) and during drought (which is usually midsummer around here and requires plants that will bloom when there is little rain). So I would focus on plants to fill those gaps. A variety of honey plants in general will tend to fill more gaps than focusing on only one or two plants. It certainly doesn't hurt to plant some sweet clover (both yellow and white as they bloom at different times) and some white Dutch clover and some birdsfoot trefoil and some borage and some anise hyssop and some tulip poplars and some black locust, but these don't tend to fill those early and late gaps, but do tend to make some honey and *may* fill a gap. Early plants that provide pollen are red maples, pussy willows, elms, crocuses, redbud, wild plums, choke cherries and other fruit trees. Dandelions are always good to have around. You can pick the dried heads from people whose lawns are full of them. Just pluck them and put them in a grocery sack and take them home and scatter them. Chicory and goldenrod often bloom in a drought and will bloom usually from about July until a killing frost. Asters are a good late blooming plant. The main thing to keep in mind, though, is that you're just trying to fill the gaps, not trying to create a crop.

Queen excluders?

The use of queen excluders has been controversial among beekeepers since the early days of their existence. I quit using them very early in my beekeeping. The bees did not want to go through them and they did not want to work the supers on the other side of them. They seemed very unnatural and constraining to me. I think they are handy to have around for things like queen rearing or a desperate attempt to find a queen, but I don't commonly use them.

The reasoning for using them:

The queen will be easier to find if I can narrow down the area I have to look. But I find the area I have to look is pretty narrow. I seldom find her other than where the highest concentration of bees is and that usually narrows it to a few frames. But this is a good reason if you need to find the queen often. In queen rearing this can be once a week or so and a queen excluder can save you some time.

Preventing brood in the supers. The only reasons I've seen a queen lay in the supers are, that she ran out of room in the brood nest, therefore she would have swarmed if she couldn't, or she wanted room to lay drones and there is no drone comb in the brood nest. Since brood comb is difficult to tear down because of cocoons, and supers usually have soft wax with no cocoons which is easily reworked the bees will build drone comb there if they don't have enough in the brood nest. If you don't want brood in the supers, give them some drone comb in the brood nest and you will have made great strides in this regard. Also, if you use all the same size box, you'll have no problem *if* she lays in the "supers" putting those frames back down in the brood nest, and if you use no chemicals, you can steal a frame of honey from there to fill out your super.

If you want to use them

If you want to use an excluder, remember you have to get the bees going through it. Using all the same sized boxes, again, will help in this regard as you can put a couple of frames of open brood above the excluder (being careful not to get the queen of course) and get them going through the excluder. When they are working the super you can put those combs back down in the brood nest. Another option (especially if you don't have the same sized boxes) is to leave out the excluder until they are working the first super and then put it in (again making sure the queen is below it and the drones have a way out the top somewhere).

"Beginning beekeepers should not attempt to use queen excluders to prevent brood in supers. However they probably should have one excluder on hand to use as an aide in either finding the queen or restricting her access to frames that the beekeeper must want to move elsewhere" -The How-To-Do-It book of Beekeeping, Richard Taylor

Queenless bees?

BLUF: Put a frame of open brood and eggs in the hive and you don't need to worry about this.

The question comes up all the time on beekeeping forums: "Are my bees queenless?" The symptoms leading to this question vary greatly and the time of year for the question varies greatly, but it is a very important question to get an answer to, or at least a resolution to and is sometimes remarkably more complex that it appears.

The most likely cause for the question is a lack of eggs and brood. Many beginning beekeepers couldn't find a queen if you marked her, clipped her and put her on one frame for them to find her, and even an experienced beekeeper in a well populated hive on a given day may have trouble finding one. So not seeing her doesn't prove anything. Not seeing eggs and brood is an important clue, but it doesn't mean that there is not a queen. It means there is not a laying queen and has not been one for a while, or you can't spot eggs. But there very well may be a virgin queen that is not laying yet.

Let's do a bit of bee math. If you accidentally kill a queen today, how long before you'll see eggs from a replacement raised by the bees? About 26 days. How much open and capped brood will there be left by the time you see eggs from the new emergency queen? The answer is none. If bees lost a queen today, and started from four-day-old larvae (four days from the egg) to raise a queen, it would be another 12 days before she emerged. Another week for her to harden and orient. And another week to get mated and start to lay. That's approximately 26 days (give or take a week). In 26 days every egg has hatched, been capped and emerged. There is now no brood left in the hive, but, in this case, there is a queen.

The problem is if the new queen flew out to mate and didn't make it back, and the hive is truly queenless, the hive looks the same. No eggs, no brood, not even any capped brood. So how do you answer the question? You give them a frame of brood with eggs and see what they do. If you have a queen cell in a couple of days, then they are queenless. You can either get a queen for them or let them raise that one.

Another problem is when you find a few eggs and a few larvae and they are very scattered. This is sometimes due to laying workers but the bees have still kept up with removing the drone eggs from the worker cells, except for a few. But what if it's a new queen that is just starting to lay? Usually she will lay in a patch and not scattered all over. Laying workers require a lot more effort to deal with.

One way to get a clue as to whether a hive is queenless is listening to it. If you don't know what a queenless hive sounds like, try catching a queen and removing her from a hive and then wait a few minutes and listen. The hive will set up a roar. This is sometimes called a "queenless roar".

Another clue that there probably is a queen who is about to start laying, is to look for a patch of empty cells surrounded by nectar, in the cluster, where they have cleared a spot for her to lay.

A grouchy hive is often a sign they are queenless or a lethargic hive. But you still need to look for eggs and larvae.

The bottom line is that queenlessness is difficult to diagnose definitively. A combination of several of these symptoms (lack of eggs and brood, queenless roar, lethargy or anger) tends to convince me. But only one or two, I give them a frame of open brood with eggs and see what happens.

Of course this illustrates why you need more than one hive.

For more information see the section *Panacea* in the Chapter *BLUF*.

Requeening

There are several questions to do with this. One is "how often should I requeen?" Beekeepers have many opinions on this ranging from twice a year to never. I tend to let them requeen themselves, but then I have a handle on swarming and I do requeen if they are too defensive or are not doing well.

The second question is "how do I requeen?" This may involve several questions such as "what do I do if I can't find the old queen?" or "how do I know they will accept the new queen?"

I have not had good luck releasing a queen if they have a queen. About the only way to do this is if you raise your own queens and you introduce a cell or a virgin queen with a lot of smoke to cover her appearance in the hive. That way it is more likely to be perceived as a supersedure by the bees. Otherwise you need to remove the old queen in order to introduce a new laying queen. If you absolutely can't find the old queen and you absolutely think you need to introduce the new one, I'd use a push in cage. All in all it's the most reliable method anyway.

A standard candy release usually works fine if there aren't any complications (such as laying workers, angry hive, already rejected a queen, been queenless a long time, can't find the old queen etc.). This is where you uncork the candy end of the cage, (or in the case of the California cages, you add the plastic tube that has the candy in it or stuff a miniature marshmallow in the hole) and you put the cage in the hive and wait for the bees to eat the candy and release the queen. It is advantageous to acceptance to release the attendants in the queen cage, but if you are a beginner you may find that intimidating. A Queen Muff (from Brushy Mt.) will help much in this as you can do all of your manipulations in a situation where the queen can't fly off on you. If you catch the queen and put her head in the cage she will usually run back in.

Putting a queen cell in anywhere the bees are thick enough to keep it warm works well.

Push In Cage

This is the most reliable release for a laying queen. The concept of this is to give the queen some newly emerged attendants, who will accept her since they have never had any other queen, some food and a place to lay. Once she is a laying queen with attendants the hive will usually accept her without protest.

Making a Push-In-Cage

Most people make these about 4 inches square (10 cm). I prefer to make them bigger. The larger they are the easier it is to get some honey (so she doesn't starve) some open cells (so she has a place to lay) and some emerging brood (so she has attendants). I like mine about 5 by 10 inches (12.5cm by 25cm). Cut some #8 hardware cloth (8 wires to an inch or $^1/_8$" wire cloth) $6^1/_2$" by $11^1/_2$" (about 16cm by 29cm). Pull off the first three wires all the way around leaving $^3/_8$" wires sticking out with no cross wires. This is to push into the comb so that the bees can't get under easily. Now come in $^3/_4$" from the corners (three more wires) and make a cut $^3/_4$" in (3 more wires) on all four corners. It really doesn't matter from which direction, but you're going to fold it around the corner. Fold the $^3/_4$" edge over. A board or the sharp edge of a table is helpful in doing this. Fold the $^3/_4$" corners over. You now have a box with no bottom that is $^3/_4$" tall and 5" by 10".

Using a Push-in-Cage

Find a comb with emerging brood. This comb is bees who are fuzzy and struggling to get out of a cell they have just chewed open. A bee with its head sticking out of a cell is emerging brood. A bee with its behind sticking out of a cell is a nurse bee feeding a larvae or a house bee cleaning a cell. Shake (if the comb is strong enough) or brush all of the bees off of the comb. Release the queen on one side of the comb where there is emerging brood and some open honey. Put the cage over her so that it has both honey and emerging brood in it. Some open cells are nice too. Push the cage into the comb. It should stick up about $^3/_8$" above the comb to make room for the queen to move around. Make room in the hive for this frame plus the $^3/_8$". Some will have enough space and some will have to have a frame removed, but you need to have the frame with the push in cage and then $^3/_8$" space between the cage and the comb on the next frame ($^3/_4$" total) so that bees have access to the cage to meet the queen and feed them if they like. Come back in four days and release the queen by removing the cage.

How do I keep queens for a few days?

If you need to keep queens that come in cages with attendants and candy, you can minimize the stress by keeping them in a cool (like 60 to 70 F) dark (like a closet) quiet (like a closet or the basement) place and give them a drop of water everyday so they can digest the candy and they will usually keep for a couple of weeks if they weren't too stressed to start with and the attendants are healthy. Give them a drop as soon as you receive them and one a day after that. If the candy looks like it will run out, you might have to give them a drop of honey and a drop of water every day. If all the attendants are dead they will need new attendants.

What's an inner cover for?

An inner cover was invented to create an air space to cut down on condensation on the cover. The original ones were made of cloth but over time the wooden ones took over. In the North the problem with winter is condensation and most of that is on the lid. The warm moist air from the cluster hits the cold lid, condenses and drips down on the cluster. An inner cover was designed to prevent this. Over the years, many other uses have been found for them. You can put an inverted jar over the hole to feed. You can put wet (just harvested and extracted) supers over them to get the bees to clean them up. You can put a porter bee escape in the hole to get the bees out of a super (I've never had much luck with this). You can double screen the hole and use it between a nuc above and a hive below in the spring or fall to help the nuc stay warm. (This has not worked well for me in the winter due to condensation).

Can I *not* use an inner cover?

If you use migratory covers, you won't need one and probably don't want one. If you use a telescopic cover it will keep the cover from getting glued down with propolis. It's difficult to remove a telescopic cover that is propolized down to the box with no inner cover as there is no where to get your hive tool in to pry it apart. If you have a telescopic cover, I recommend you use the inner cover. If you live in the north and want to use migratory covers, make sure there is some kind of top entrance (you can cut a notch in the cover to make one. See Brushy Mt. migratory covers for an example) and put some Styrofoam on top of the lid with a brick on top of the Styrofoam. The Styrofoam will keep the lid from being as cold and the vent at the top (through the notch) will allow the moist air out.

What's that smell?

Smells are always best investigated. They are very subjective and therefore it's best for you to see it for yourself to associate that smell with that occurrence. The most common smell that people get worried about is the smell of goldenrod honey ripening. This happens sometimes between summer and fall. To me, it smells like old gym socks. Some people say it smells like butterscotch. Most people think it smells sour.

If you smell the smell of rotting meat, I would investigate. Sometimes you have piles of dead bees from a pesticide kill or robbing. Sometimes you have a brood disease. It's worth investigating to see what the cause is.

What's the best beekeeping book?

All of them. Read every beekeeping book you can get your hands on. But my favorites are the old ABC XYZ of Bee Culture, Langstroth's Hive and the Honey Bee, everything by Richard Taylor and Brother Adam and the ones that I've posted on my classic bee books page. (http://www.bushfarms.com/beesoldbooks.htm) In addition if you're past all the beekeeping books and want to know even more, all of Eva Crane's books are fascinating.

For a beginner's book for natural beekeeping, *The Complete Idiots Guide to Beekeeping* is awesome. For general beginning beekeeping, *Backyard Beekeeping* by Kim Flottum is very good and simple.

What's the best breed of bees?

There has been much speculation by beekeepers for many centuries on this. I suppose at the turn of the 19th to the 20th century there was probably the most agreement. Italians were pretty much what everyone wanted. Now there are just as many who want Carniolans or Caucasians or Buckfasts or Russians. I see more variation from hive to hive than race to race. I'd say the best breeds of bees are the ones that are surviving around you. That's what I'm raising.

But if you want to buy some queens, the issues are how well they do in your climate (for instance Italians are probably better adapted to the South and Carniolans are better adapted to the North), and health (hygienic behavior, tracheal mite resistance, Varroa mite resistance etc.).

Why are there all these bees in the air?

Another panicked posting on the bee forums several times a year will involve a lot of bees flying. This is usually interpreted by the new beekeeper as either a swarm or robbing. A swarm does put a lot of bees in the air, but they are going somewhere. In this case they are just hovering around the hive. If the bees seem happy and organized and not frantic and fighting on the landing board, and especially if it's short-lived and on a sunny afternoon; then it's probably just young bees orienting for the first time. Look for signs of wrestling or fighting on the landing board to rule out robbing. If there are no signs of robbing, this is the sign of a healthy hive. If the hovering bees seem to be leaving a trail of bees as they fly off, then it's probably a swarm gathering in one of your trees.

Why are there bees on the outside of my hive?

Typically beekeepers call this bearding because it often looks like the hive has a beard. Causes are heat, congestion and lack of ventilation. Make sure they have room and ventilation and don't worry about it.

Bees bearding is like people sweating. It's what bees do when they are hot.

It's good to cover the bases and then accept it. If you were sweating you'd take what steps were reasonable (turn on the fan, open the window, take off your sweater, drink lots of water) and then you'd accept that it's just hot.

With the bees, make sure they have top and bottom ventilation, (open the bottom entrance, remove the tray if you have a SBB, prop open the top box, slide a super back to make a gap) make sure they have enough room (put supers on as needed) and then don't worry about it. Bearding is not proof they are about to swarm. It is proof they are hot. I think lack of ventilation contributes to an "overcrowding swarm" but it's not the only cause and it's nothing to be concerned about if you've taken care of the bees having ventilation and room.

Why are they dancing at the entrance in unison?

A few times a year some new beekeeper wants to know what the bees are doing line dancing (rhythmically swaying) on the landing board. This is called "washboarding" and actually no one knows why they do it, but they do. Personally I think it's a social dance. Perhaps even a thanksgiving dance.

Why not use an electric fan for ventilation?

The subject comes up a lot. I've never quite understood it, but I supposed it comes back to a desire to "help". Bees, however, have a very efficient and precise ventilation system and anything you do will probably interfere with it rather than help. The problem with an electric one is that the bees will find themselves fighting the ventilator. I think you're much better off to just give them some ventilation top and bottom and let them control it.

Why did my bees die?

With a death over winter, a post mortem would be to check:
- Are they not in contact with stores? It doesn't matter if they have honey if they can't get to it because they are stuck. If they are not in contact with stores they starved.
- If they are in contact with stores, are there thousands of dead Varroa on the bottom board or the tray under the SBB (I would have it in, of course)? If so, I think it's safe to say the primary cause was Varroa.
- Are there a lot of little clusters of bees in the hive instead of one large cluster? If so I would suspect Tracheal mites.
- Are the bees wet and moldy? If so I would suspect condensation got them wet and wet bees seldom survive.
- It is a commonly held belief that bees dead headfirst in the cells means they starved. All dead hives over winter will have many bees with their heads in cells. That's how they cluster tightly for warmth. I would

read more into whether or not they are on contact with stores.

- With a death during the active season, I'd look for piles of dead bees and if there are signs of robbing. Robbing can lead to piles of dead bees, but there are other symptoms like ragged comb and frantic bees. Pesticides usually have crawling dying bees and piles of them dead. A dwindling hive, you should probably check the brood on to make sure you don't have a brood disease.

Why do bees make different colors of wax?

Bees only produce one color of wax—white.

If they track a lot of pollen on the wax it turns yellow. If they raise brood in it, it turns brown from the cocoons. If they leave enough cocoons, it turns black.

As far as cappings, they produce two kinds. On honey it is made of wax which is air tight to keep the honey from absorbing moisture, so it starts white until they track pollen on it which may turn it yellow. On brood it is a mixture of wax and cocoons which can breathe so the pupae can get oxygen. Depending on how old and dark the cocoons are and how much are available, they vary from light yellow to dark brown.

How often should I inspect?

If you are a new beekeeper you should inspect often. Not because the bees need you to, but because you can't learn anything if you don't observe. As far as the bees are concerned you only need to check often enough to not let them run out of room. How often? Well I would try not to totally disrupt them every day. If you have an observation hive you can learn a lot there. If you have a window on the hive or a Plexiglas inner cover you can observe more. But with a typical hive I would figure on opening the hive once a week or so until you are comfortable guessing what is going on inside by assessing the outside. Eventually, if you think about what you expect to see and open it and see if you're right, you'll get good at assessing without opening.

Should I drill a hole?

Usually the idea is either for a top entrance or for ventilation. I don't like holes in my equipment. Here are times I regretted drilling holes:
- Times I wanted to close up a hive and forgot the hole. (moving and using a bee escape come to mind)
- Times I accidentally put my hand either over, under or in the hole when lifting the super.
- Times in winter when I wanted to close it up more.
- Times that a hive gets weak and forgets to guard both entrances and they get robbed and I have to find a way to close it off.
- Times that I need a box without a hole and the only one handy has a hole in it.

There is nothing you accomplish by putting a hole in the box that you can't do by sliding the box back $^3/_4"$ or putting in a couple of shingle shims or using a Imirie shim.

If you have holes in your equipment you can plug them with a tin can lid tacked over the hole. In the beeyard in a pinch you can plug them temporarily with a wad of beeswax.

How do you brush bees?

There are two primary ways to get bees off combs. Brushing and shaking. Practice a few different techniques to see what works for you shaking them. It will depend on many things. New soft comb (on foundation or not, wired or not) that is heavy with honey will break if you shake it too hard. When hot it is even softer. Foundationless that is not attached all the way around will be even more fragile. These should be brushed. Old black brood comb will not break no matter how hard you slam it. Older comb that is not so soft you can shake well enough without breaking it, but there is a limit and you need to learn that limit based on all of the variables (new, soft, old full of cocoons, heavy with honey, light with brood etc.). Also don't shake a frame with queen cells or you'll damage the queen. Use a brush. Doing a double shake (one shake immediately followed by a second as fast as you can) works if you do it just right. Practice it until you get it to work. You can "pound" bees as C.C. Miller called it. You grab the end of the top bar firmly and hit your other fist on that fist. The jar will knock them off.

It's one of those things that is more art than science but there are principles, and the primary one is surprise. The secondary one is that it's hard, not soft. It seems contrary because normally in beekeeping you are trying to be slow and graceful and not do anything suddenly. And to get bees off you have to be sudden and hard. There is no graceful and soft way to do it.

How many cells on a frame?

Deep frame of 5.4mm foundation 7000
Deep frame of 4.9mm foundation 8400
Medium frame of 5.4mm foundation 4620

Medium frame of 4.9mm foundation 5544

Burr comb?

The main cause of burr between boxes is thin top bars. Plastic frames all have these. I just accept it.

> *"...that very practical Canadian bee-keeper, J.B. Hall, showed me his thick top-bars, and told me that they prevented the building up of so much burr-comb between the top-bars and the sections...and I am very glad that at the present day it can be dispensed with by having top-bars 1-$^1/_8$ inch wide and $^7/_8$ inch thick, with a space of $^1/_4$ inch between top-bar and section. Not that there is an entire absence of burr-combs, but near enough to it so that one can get along much more comfortably than with the slat honey-board. At any rate there is no longer the killing of bees that there was every day the dauby honey-board was replaced."--C.C. Miller, Fifty Years Among the Bees.*

> *"Q. Do you believe that a half-inch thick brood-frame top-bar will tend to prevent the bees building burr-comb on such frames, as well as the three-quarter inch top-bar? Which kind do you use?*

A. I do not believe that the one-half inch will prevent burr-combs quite as well as the three-quarter. Mine are seven-eighths."--C.C. Miller, A Thousand Answers to Beekeeping Questions

Appendix to Volume I: Glossary

Note: many of these terms are Latin and the plural of the ones with an "a" ending will be "ae". The plural of the "us" endings will be "I". Also meanings are given in the context of beekeeping.

7/11 or Seven/Eleven = Foundation with a cell size that is 700 cells per square decimeter with 11 cells left over. Hence 7/11. Actually 5.6mm cell size. Used because it is a size the queen dislikes laying in because it's too big for worker brood and too small for drone brood. If the queen does lay in it, it will usually be drones. It's only currently available from Walter T. Kelley

A

Acute Paralysis Virus aka APV = A viral disease of adult bees which affects their ability to use legs or wings normally. It can kill adults and brood.

Abdomen = The posterior or third region of the body of the bee that encloses the honey stomach, stomach, intestines, sting and the reproductive organs.

Abscond = When the entire colony of bees abandons the hive because of pests, disease or other adverse conditions.

Acarapis woodi = Tracheal Mite, which infests the bees' trachea; sometimes called Acarine Disease or Isle of Wight disease.

Acarapis dorsalis = Mite that lives on honey bees that is indistinguishable from Tracheal mites (Acarapis woodi). It is classified differently simply based on the location where it is found.

Acarapis externus = Mite that lives on honey bees that is indistinguishable from Tracheal mites (Acarapis woodi). It is classified differently simply based on the location where it is found.

Accelerated queen rearing = A system of mating nucs where there are usually two queens in the mating nuc a week apart, one in a nursery cage and one loose and mating. Every week the one that is now mated is removed the one in the cage is released and the new cell is put in with a hair curler cage on it.

Africanized Honey Bees = I have heard these called Apis mellifera scutelata But Scutelata are actually African bees from the Cape. They used to be called Adansonii, at least that's what Dr. Kerr, who bred them, thought they were. AHB are a mixture of African (Scutelata) and Italian bees. They were created in an attempt to increase production of bees. The USDA bred these at Baton Rouge from stock obtained from Dr. Kerr in Brazil. The USDA shipped these queens to the continental US over the course of many years. The Brazilians also were experimenting with them and the migration of those bees has been followed in the news for some time. They are extremely productive bees that are extremely defensive. If you have a hive hot enough that you think they are AHB you need to requeen them. Having angry bees where they might hurt people is irresponsible. You should try to requeen them (see the chapter *Requeening a hot hive* in Volume 3) so no one (including you) gets hurt.

Afterswarm = A swarm after the primary swarm. These are headed by a virgin queen.

Alarm pheromone = A chemical (iso-pentyl acetate) substance which smells similar to artificial banana flavoring, released near the worker bee's sting, which alerts the hive to an attack.

Alcohol wash = Putting a cupful of bees in a jar with alcohol to kill the bees and mites so you can count the Varroa mites. A sugar roll is a non-lethal method of doing the same.

Allergic reaction = A systemic reaction to something, such as bee venom, characterized by hives, breathing difficulty, or loss of consciousness. This should be distinguished from a normal reaction to bee venom, which is itching and burning in the general vicinity of the sting.

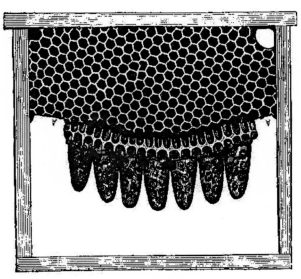

Alley Method

Alley Method = A graftless method of queen rearing system where bees are put in a "swarm box" to convince them of their queenlessness and a strip of old brood comb is cut and put on a bar for the bees to build into queen cells.

American Foulbrood = For more detail see the chapter on *Enemies of the Bees.* Caused by a spore forming bacteria. It used to be called Bacillus larvae but has recently been renamed Paenibacillus larvae. With American Foul Brood the larvae usually dies after it is capped, but it looks sick before. The brood pattern will be spotty. Cappings will be sunken and sometimes pierced. Recently dead larvae will string when poked with a match-stick. The smell is rotten and distinctive. Older dead larvae turn to a scale that the bees cannot remove.

Anaphylactic shock = Constriction of the smooth muscle including the bronchial tubes and blood vessels of a human, caused, in the context of beekeeping, by hypersensitivity to venom possibly resulting in sudden death unless immediate medical attention is received.

Antenna = One of two sensory organs located on the head of the bee, which enable bees to smell and taste.

Attendants = Worker bees that are attending the queen. When used in the context of queens in cages, the workers that are added to the cage to care for the queen.

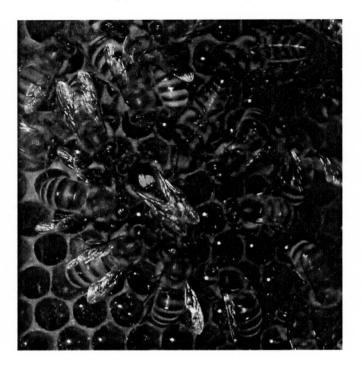

Apiary = A bee yard.

Apiarist = A beekeeper.

Apiculture = The science and art of raising honey bees.

Apis mellifera mellifera = These are the bees
native to England or Germany. They have some of the
characteristics of the other dark bees. They tend toward
being runny (excitable on the combs) and a bit swarmy,
but also seem to be well adapted to damp Northern
climates.

Apis mellifera = Includes the honey bees originating in Africa and Europe.

B

Bacillus larvae = The outdated name for Paeni-bacillus Larvae, the bacteria that causes American Foulbrood.

Bacillus thuringiensis = A naturally occurring bacteria that is sprayed on empty comb to kill wax moths. Also sold to control larvae of other specific insects.

Backfilling = A term coined by Walt Wright to describe the process of the bees creating a honey bound brood nest. The process where the bees put honey in the brood nest to prevent the queen from laying to prepare for swarming.

Baggie feeder = These are just gallon zip lock baggies that are filled with three quarts of syrup, laid on the top bars and slit on top with a razor blade with two or three small slits. The bees suck down the syrup until the bag is empty. A box of some kind is required to make room. An upside down miller feeder or a one by three shim or just any empty super will work. Advantages are the cost (just the cost of the bags) and the bees will work it in cooler weather as the cluster keeps it warm. Disadvantages are you have to disrupt the bees to put new bags on and the old bags are ruined.

Bait Hive aka Decoy hive aka Swarm trap = A hive placed to attract stray swarms. Optimum bait hive: At least 20 liters of volume. 9 feet off the ground. Small entrance. Old comb. Lemongrass oil. Queen substance.

Balling = Worker bees surrounding a queen either to confine her because they reject her or to confine her to protect her.

Banking queens = Putting multiple caged queens in a nuc or hive.

Bearding = When bees congregate on the front of the hive.

Bee blower = A gas or electrically driven blower used to blow bees from supers when harvesting.

Bee bread = Fermented pollen stored in the hive to use to feed brood.

Bee brush = Soft brush or whisk or large feather or handful of grass used to remove bees from combs.

Bee escape = A device constructed to permit bees to pass one way, but prevent their return; used to clear bees from supers or other uses. The most common one seems to be the Porter escape which is made to go in the hole in the inner cover. The most effective one seems to be the triangular one which is its own board.

Bee Go = Butyric which is used to drive bees from supers. This smells a lot like vomit.

Bee Gum = A piece of a hollow tree used for a hive.

Bee haver = A term coined by George Imirie. One who has bees but has not learned enough technique to be a beekeeper.

Bee jacket = A white jacket, usually with a zip on veil and elastic at the sleeves and waist, worn as protection when working bees.

Bee Parasitic Mite Syndrome aka Parasitic Mite Syndrome = A set of symptoms that are caused by a major infestation of Varroa mites. Symptoms include the presence of Varroa mites, the presence of various brood diseases with symptoms similar to that of foulbroods and sacbrood but with no predominant pathogen, AFB-like symptoms, spotty brood pattern, increased supersedure of queens, bees crawling on the ground, and a low adult bee population.

Bee Quick = A chemical, that smells like benzal-dehyde that is used to drive bees from supers.

Bee space = A space between $^1/_4$ and $^3/_8$ inch which permits free passage for a bee but too small to encourage comb building, and too large to induce pro-polizing.

Bee suit = A pair of white coveralls made for beekeepers to protect them from stings and keep their clothes clean. Most come with zip-on veils.

Bee tree = A hollow tree occupied by a colony of bees.

Bee vac aka Bee vacuum = A vacuum used to suck up bees when doing a cutout or removal. Usually converted from a shop vac. It needs careful adjustment to not kill the bees.

Bee veil = Netting or screen for protecting the beekeeper's head and neck from stings.

Bee venom = The poison secreted by special glands attached to the stinger of the bee which is injected into the victim of a sting.

Beehive = A box usually with movable frames, used for housing a colony of bees.

Beelining = Finding feral bees by establishing the line which the bees fly back to their home. This can also include marking and timing the bees to get the distance and triangulating the location by releasing the bees from various places.

Beek = Beekeeper

Beekeeper = One who keeps bees. An Apiarist.

Beeswax = A substance that is secreted by bees by special glands on the underside of the abdomen, deposited as thin scales, and used after mastication and mixture with the secretion of the salivary glands for constructing the honeycomb. The melting point of beeswax is 144 to 147 °F.

Better Queens method = A graftless queen rearing method similar to Isaac Hopkins' actual queen rearing method (as opposed to the "Hopkins Method"). Sort of the Alley Method but with new comb instead of old.

Betterbee = A beekeeping supply company out of New York. They have many things no one else does. They also have eight frame equipment.

Benzaldehyde = A colorless nontoxic liquid alde-hyde C6H5CHO that has an odor like that of bitter almond oil, that occurs in many essential oils and is sometimes used to drive bees out of honey supers. Also the flavor added to Maraschino cherries. What Bee Quick smells like.

Black scale = Refers to dried pupa, which died of American Foulbrood.

Boardman feeder = These come in all the be-ginners' kits. They go in the entrance and hold an in-verted quart mason jar. I'd keep the jar lid and throw away the feeder. They are notorious for causing rob-bing. They are easy to check but you have to shake off the bees and open the jar to refill them.

Bottling tank = A food grade tank holding 5 or more gallons of honey and equipped with a honey gate to fill honey jars.

Bottom bar = The horizontal piece of the frame that is on the bottom of the frame.

Bottom board = The floor of a bee hive.

Bottom board feeder = This is picture of the bottom board feeder that Jay Smith came up with. It's simply a dam made with a $^3/_4$″ by $^3/_4$″ block of wood put an inch or so back from the where the front of the hive would be (18″ or so forward of the very back). The box is slid forward enough to make a gap at the back. The syrup is poured in the back. A small board can be used to block the opening in the back. The bees can still get out the front by simply coming down forward of the dam. The picture is from the perspective of standing behind the hive looking toward the front. The edges of the dam have been enhanced and labels put on to try to make more sense. This version doesn't work on a weak hive as the syrup is too close to the entrance. It drowns as many bees as the frame feeders.

Bottom supering = The act of placing honey supers under all the existing supers, directly on top of the brood box. The theory is the bees will work it better when it's directly above the brood chamber; as opposed to *top* supering which would be just putting the supers on top of the existing supers.

Box Jig = Jig for nailing boxes. (for more pictures see chapter by that name in Volume 3)

Brace comb = A bit of comb built between two combs to fasten them together, between a comb and adjacent wood, or between two wooden parts such as top bars.

Braula coeca = A wingless fly commonly known as the bee louse.

Breeder hive = The hive from which eggs or larvae are taken for queen rearing. In other words the donor hive.

Bricks = Used to keep the lids from blowing off in the wind and often used in particular configurations as visual clues as to the state of a hive.

Brood = Immature bees not yet emerged from their cells; in other words, egg, larvae or pupae.

Brood chamber = The part of the hive in which the brood is reared; may include one or more hive bodies and the combs within. Sometimes used to refer to a deep box as these are commonly used for brood.

Brood nest = The part of the hive interior in which brood is reared; usually the two bottom boxes.

Brushy Mountain = A beekeeping supply company out of North Carolina. A big proponent of all mediums and eight frame boxes. They have many items no one else has.

Bt = Bacillus thuringiensis. A naturally occurring bacteria that is sprayed on empty comb to kill wax moths. Also sold to control larvae of other specific insects.

Buckfast = A strain of bees developed by Brother Adam at Buckfast Abbey in England, bred for disease resistance, disinclination to swarm, hardiness, comb building and good temper.

Burr comb = Small pieces of comb outside of the normal space in the frame where comb usually is. Brace comb would fall into this category.

C

Candy plug = A fondant type candy placed in one end of a queen cage to delay her release.

Capped brood = Immature bees whose cells have been sealed over with papery caps.

Capping melter = Melter used to liquefy the wax from cappings as they are removed from honey combs.

Cappings = The thin wax covering over honey; once cut off of extracting frames.

Capping scratcher = A fork-like device used to remove wax cappings covering honey, so it can be extracted. Usually used on low areas that get missed by the uncapping knife.

Carniolan bees = Apis mellifera carnica. These are darker brown to black. They fly in slightly cooler weather and in theory are better in northern climates. They are reputed by some to be less productive than Italians, but I have not had that experience. The ones I have had were very productive and very frugal for the winter. They winter in small clusters and shut down brood rearing when there are dearths.

Castes = The three types of bees that comprise the adult population of a honey bee colony: workers, drones, and queen

Carts = Used for wheeling boxes or hives around.

Caucasian bees = Apis mellifera caucasica. They are a silver gray to dark brown color. They do propolis excessively. It is a sticky propolis rather than a hard propolis. They often coat everything with this sticky kind of proplolis, like fly paper. They build up a little slower in the spring than the Italians. They are reputed to be more gentle than the Italians. Less prone to robbing. In theory they are less productive than Italians. I think on the average they are about the same productivity as the Italians, but since they rob less you get less of the really booming hives that have robbed out all their neighbors.

Cell = The hexagonal compartment of a honey comb.

Cell bar = A wooden strip on which queen cups are suspended for rearing queen bees.

Cell cup = Base of an artificial queen cell, made of beeswax or plastic and used for rearing queen bees or an empty beginning of a queen cell that the bees often build for no reason.

Cell finisher = A hive used to finish queen cells i.e. take them from capped to just before emergence. Sometimes queenright, sometimes queenless.

Cell starter = A hive used to start queen cells i.e. take them from just grafted to capped. Sometimes a "swarm box" or sometimes just a queenless hive.

Chalkbrood = This is caused by a fungus Ascosphaera apis. It arrived in the US in 1968. If you find white pellets in front of the hive that kind of look like small corn kernels, you probably have chalkbrood. Putting the hive in full sun and adding more ventilation usually clears this up. Honey instead of syrup may contribute to clearing this up, since sugar syrup is much more alkali (higher pH) than honey.

Checkerboarding (aka Nectar Management) = A method of swarm control and hive management pioneered by Walt Wright, that involves putting alternating frames of capped honey and empty drawn comb above the brood nest in late winter.

Chest hive = a hive that is laid out horizontally instead of vertically.

Chilled brood = Immature bees that have died from exposure to cold; commonly caused by mismanagement or sudden cold spells.

Chimney = When the bees fill only the center frames of honey supers.

Chinese grafting tool = Grafting tool made of plastic, horn and bamboo that has a retractable "tongue" that slides under the larvae and, when released, pushes it off of the "tongue". Popular because it is easier to operate than most grafting needles and it lifts up more royal jelly in the process. Quality varies and most recommend buying several and picking the ones you like out of those.

Chitin = Material which the exoskeleton of an insect is made of.

Chronic Paralysis Virus aka CPV = Symptoms: bees trembling, unable to fly, with K-wings and distended abdomens. One variety called the hairless black syndrome, is recognized by hairless, black shiny bees crawling at the hive entrance.

Chunk honey = Cut comb honey packed into jars then filled with liquid honey.

Clarifying = Removing visible foreign material from honey or wax to increase its purity.

Clipping = The practice of taking part of one or both wings off of a queen both for discouraging or slowing swarming and for identification of the queen.

Cloake Board AKA FWOF (Floor without a floor) = A device to divide a colony into a queenless cell starter and reunite it as a queenright cell finisher without having to open the hive.

Cloake board

Cluster = The thickest part of the bees on a warm day, usually the core of the brood nest. On a day below 50º F the only location where the bees are. It is used to refer both to the location and to the bees in that location.

Cocoon = A thin silk covering secreted by larval honey bees in their cells in preparation for pupation.

Coffin hive = a hive that is laid out horizontally instead of vertically.

Colony = The superorganism made up of worker bees, drones, queen, and developing brood living together as a family unit.

Colony Collapse Disorder = A recently named problem where most of the bees in most of the hives in an apiary disappear leaving a queen, healthy brood and only a few bees in the hive with plenty of stores.

Comb = The wax structures in a colony in which eggs are laid, and honey and pollen are stored. Shaped like hexagons.

Comb foundation = A commercially made structure consisting of thin sheets of beeswax with the cell bases of a particular cell size embossed on both sides to induce the bees to build a that size of cell.

Comb Honey = Honey in the wax combs, either cut from larger combs or produced and sold as a separate unit, such as a wooden section $4^1/_2$" square, or a plastic round ring.

Conical escape = A cone-shaped bee escape, which permits bees, a one-way exit; used in a special escape board to free honey supers of bees.

Cordovan bees = A subset of the Italians. In theory you could have a Cordovan in any breed, since it's technically just a color, but the ones for sale in North American that I've seen are all Italians. They are slightly more gentle, slightly more likely to rob and quite striking to look at. They have no black on them and look very yellow at first sight. Looking closely you see that where the Italians normally have black legs and head, they have a purplish brown legs and head.

Creamed honey = Honey that has undergone controlled granulation to produce a finely textured candied or crystallized honey which spreads easily at room temperature. This usually involves adding fine "seed" crystals and keeping at 57º F (14º C).

Crimp-wired foundation = Comb foundation in-to which crimp wire is embedded vertically during foun-dation manufacture.

Crimper = A device used to put a ripple in the frame wire to both make it tight and to distribute stress better and give more surface to bind it to the wax.

Cupralarva = A particular brand of graftless queen rearing system.

Cut-comb Honey = Comb honey cut into various sizes, the edges drained, and the pieces wrapped or packed individually

Cut-out = Removing a colony of bees from somewhere that they don't have movable comb by cutting out the combs and tying them into frames.

D

Dadant = A beekeeping supply company out of Illinois. Founded by C.P. Dadant who was a pioneer in the modern beekeeping era and invented, among other things, the Jumbo and the square Dadant box. ($19^7/_8$" by $19^7/_8$" by $11^5/_8$"), published and wrote for the American Bee Journal and translated *Huber's Observations on Bees* from French to English and published many books including but not limited to the later versions of *The Hive and the Honey Bee*.

Dadant deep = A box designed by C.P. Dadant that is $11\ ^5/_8$" deep and the frame is $11^1/_4$" deep. Sometimes called Jumbo or Extra Deep.

Dearth = A period of time when there is no available forage for bees, due to weather conditions (rain, drought) or time of year.

Decoy hive aka Bait hive aka Swarm trap = A hive placed to attract stray swarms.

Deep = In Langstroth terms, a box that is $9^5/_8$" deep and the frame is $9^1/_4$" deep. Sometimes called a Langstroth Deep.

Deformed Wing Virus = A virus spread by the Varroa mite that causes crumpled looking wings on fuzzy newly emerged bees.

Demaree = The method of swarm control that separates the queen from most of the brood within the same hive and causes them to raise another queen with the goal of a two queen hive, increased production and reduced swarming.

Depth = The vertical measurement of a box or frame.

Dequeen = To remove a queen from a colony. Usually done before requeening, or as a help for brood diseases or pests.

Detritus = Wax scales and debris that sometimes build up at the bottom of a natural colony.

Dextrose = Also known as glucose, it is a simple sugar (or monosaccharide) and is one of the two main sugars found in honey; forms most of the solid phase in granulated honey.

Diastase = A starch digesting enzyme in honey adversely affected by heat; used in some countries to test quality and heating history of stored honey.

Diploid = Possessing pairs of genes, as workers and queens do, as opposed to haploid, possessing single genes as drones do.

Disease resistance = The ability of an organism to avoid a particular disease; primarily due to genetic immunity or avoidance behavior.

Dividing = Separating a colony to form two or more colonies. AKA a split

Division = Separating a colony to form two or more colonies.

Division board = A wooden or plastic piece like a frame but tight all the way around used to divide one box into more compartments for nucs.

Division board feeder or Frame feeder = A wooden or plastic compartment which is hung in a hive like a frame and contains sugar syrup to feed bees. The original designation (Division) was because it was *used* to make a division between two halves of a box to divide it into nucs, usually for queen rearing or making increase (splits). Most of them have a beespace around them now and cannot be used to make a division.

Domestic = Bees that live in a manmade hive. Since all bees are pretty much wild this is a relative term.

Doolittle method = A method of queen rearing that involves grafting young larvae into queen cups. First discovered by Nichel Jacob in 1568, then written about by Schirach in 1767 and then Huber in 1794 and finally popularized by G.M. Doolittle in his book *Scientific Queen Rearing* in 1846.

Double screen = A wooden frame, $^{1}/_{2}$ to $^{3}/_{4}$" thick, with two layers of wire screen to separate two colonies within the same hive, one above the other. Often an entrance is cut on the upper side and placed to the rear of the hive for the upper colony and sometimes other openings are incorporated which would them be a Snelgrove board.

Double story or Double deeps = Referring to a beehive wintering in two deep boxes.

Double wide = A box that is twice as wide as a ten frame box. $32^{1}/_{2}$" wide.

Drawn combs = Full depth comb ready for brood or nectar with the cell walls drawn out by the bees, completing the comb as opposed to foundation that has not been worked by the bees and has no cell walls yet.

Drifting = The movement of bees that have lost their location and enter hives other than their own home. This happens often when hives are placed in long straight rows where returning foragers from the center hives tend to drift to the row ends or when making splits and the field bees drift back to the original hive.

> *"The percentage of foragers originating from different colonies within the apiary ranged from 32 to 63 percent"—from a paper, published in 1991 by Walter Boylan-Pett and Roger Hoopingarner in Acta Horticulturae 288, 6th Pollination Symposium (see Jan 2010 edition of Bee Culture, 36)*

Drone = The male honeybee which comes from an unfertilized egg (and is therefore haploid) laid by a queen or less commonly, a laying worker.

Drone comb = Comb that is made up of cells larger than worker brood, usually in the range of 5.9 to

7.0mm in which drones are reared and honey and pollen are stored.

Drone brood = Brood, which matures into drones, reared in cells larger than worker brood. It is noticeably larger than worker brood and the cappings are distinctly dome shaped.

Drone Congregation Area = A place that drones from many surrounding hives congregate and wait for a queen to come. In other words a mating area. Drones find them by following both pheromone trails and topographical features of the landscape such as tree rows.

Drone layers = A drone laying queen (one with no sperm left to fertilize eggs) or laying workers.

Drone laying queen = A queen that can lay only unfertilized eggs, due to age, improper or late mating, disease or injury.

Drone mother hive = The hive which is encouraged to raise a lot of drones to improve the drone side of mating queens. Based on the myth that you can make bees raise more drones. Taking drone comb from the ones you want to perpetuate and giving them to other colonies is the only real way to succeed at this as the mother colony will then raise more drones while the colonies receiving the drone comb will raise less of their own because they will be raising the ones from the drone mother.

Drumming = Tapping or thumping on the sides of a hive to make the bees ascend into another hive placed over it or to drive them out of a tree or house.

This will not get all of them out, but will move a significant number.

Dorsal-Ventral Abdominal Vibrations dance = A dance used to recruit forages. Also used on queen cells about to emerge and possibly other times.

Dwindling = Any rapid decline in the population of the hive. The rapid dying off of old bees in the spring; sometimes called spring dwindling or disappearing disease.

Dysentery = A condition of adult bees characterized by severe diarrhea (as evidenced by brown or yellow streaks on the front of the hive) and usually caused by long confinement (from either cold or beekeeper manipulation), starvation, low-quality food, or Nosema infection.

E

Eight frame = Boxes that were made to take eight frames. Usually between $13^1/_2$" and 14" wide depending on the manufacturer. Typically $13^3/_4$" wide.

Eggs = The first phase in the bee life cycle, usually laid by the queen, is the cylindrical egg $^1/_{16}$" (1.6 mm) long; it is enclosed with a flexible shell or chorion. It resembles a small grain of rice.

Eke = The term originated with skeps and it was "an enlargement" which is the equivalent of today's super. In current usage it usually refers to a shim that is either added to the top for feeding things like pollen patties or added under a shallow to make it into a deep. The term is used more frequently in Britain.

Electric embedder = A device that heats the foundation wire by running current through it for embedding of wires in foundation.

End bar = The piece of a frame that is on the ends of the frame i.e. the vertical pieces of the frame.

Entrance reducer = A wooden strip used to regulate the size of the entrance.

Escape board = A board having one or more bee escapes in it used to remove bees from supers.

European Foulbrood = Caused by a bacteria. It used to be called Streptococcus pluton but has now been renamed Melissococcus pluton. European Foul Brood is a brood disease. With EFB the larvae turn brown and their trachea is even darker brown. Don't confuse this with larvae being fed dark honey. It's not just the food that is brown. Look for the trachea. When it's worse, the brood will be dead and maybe black and maybe sunk cappings, but usually the brood dies before they are capped. The cappings in the brood nest will be scattered, not solid, because they have been removing the dead larvae. To differentiate this from AFB use a stick and poke a diseased larvae and pull it out. The AFB will "string" two or three inches.

Ether wash = Putting a cupful of bees in a jar with a spray of starter fluid to kill the bees and mites so you can count the Varroa mites. A sugar roll is a non-lethal and much less flammable method of doing the same.

European Honey Bees = Bees from Europe as opposed to bees originating in Africa or other parts of the world or bees crossbred with those from Africa.

Eyelets = Optional small metal piece fitting into the wire-holes of a frame's end bar; used to keep the reinforcing wires from cutting into the wood. Many people use a staple across where it would split the wood instead.

Extra shallow = A box that is $4^{11}/_{16}$ or $4^3/_4"$ deep. Usually used for cut comb. Sometimes modified for sections.

Extracted honey = Honey removed from combs usually by means of a centrifugal force (an extractor) in order to leave the combs intact but with hobbyists often from crushing the comb and straining it (see Crush and Strain).

Ezi Queen = A particular brand of graftless queen rearing system.

F

Frame feeder or Division board feeder = A wooden or plastic compartment which is hung in a hive like a frame and contains sugar syrup to feed bees. The original designation (Division) was because it was *used* to make a division between two halves of a box to divide it into nucs, usually for queen rearing or making increase (splits). Most of them have a beespace around them now and cannot be used to make a division.

Feeders = Any device used to feed bees.

Fermenting honey = Honey which contains too much water (greater than 20%) in which yeast has grown and caused some of it to turn into carbon dioxide, water and alcohol.

Feral (queen or bees) = Since all North American bees are considered to have come from domestic stock, what most people call "wild" bees are really "feral" bees. Some use the term for survivor bees that were captured and used to raise queens meaning they *were* feral as opposed to *are* feral.

Fertile queen = An inseminated queen.

Fertilized = Usually refers to eggs laid by a queen bee, they are fertilized with sperm stored in the queen's spermatheca, in the process of being laid. These develop into workers or queens.

Festooning = The activity of young bees, engorged with honey, hanging on to each other usually to secrete beeswax but also in bearding and swarming..

Field bees = Worker bees which are usually 21 or more days old and work outside to collect nectar, pollen, water and propolis; also called foragers.

Flash heater = A device for heating honey very rapidly to prevent it from being damaged by sustained periods of high temperature

Flight path = Usually refers to the direction bees fly leaving their colony; if obstructed, may cause bees to accidentally collide with the person obstructing and eventually become aggravated.

Floor Without a Floor AKA FWOF AKA Cloake Board = A device to divide a colony into a queenless cell starter and reunite it as a queenright cell finisher without having to open the hive.

Follower board = A thin board used in place of a frame usually when there are fewer than the normal number of frames in a hive. This is usually referring to one that has a beespace around it and is used to make the frames easier to remove without rolling and to cut down on condensation on the walls. Sometimes it's used to refer to a board that is bee tight and used to divide a box into two colonies. When designed and used in this manner it should be called a division board.

Food chamber = A hive body filled with honey for winter stores. Typically a third deep used in unlimited brood nest management.

Forage = Natural food source of bees (nectar and pollen) from wild and cultivated flowers. Or the act of gathering that food.

Foragers = Worker bees which are usually 21 or more days old and work outside to collect nectar, pollen, water and propolis; also called field bees.

Foundation = Thin sheets of beeswax embossed or stamped with the base of a worker (or rarely drone) cells on which bees will construct a complete comb (called drawn comb); also referred to as comb foundation, it comes wired or unwired and also in plastic as well as one piece foundations and frames as well as different thicknesses (thin surplus, surplus, medium) and different cell sizes (brood =5.4mm, small cell = 4.9mm, drone=6.6mm).

Foundationless = A frame with some kind of comb guide that is used without foundation.

Frame = A rectangular structure of wood designed to hold honey comb, consisting of a top bar, two end bars, and a bottom bar; usually spaced a bee-space apart in the super.

Frame feeder = Sometimes called a "division board feeder". It takes the place of one or more frames. Less bees drown if you put floats in.

Fructose = Fruit sugar, also called levulose (left handed sugar), a monosaccharide commonly found in honey that is slow to granulate

Fumagilin-B = Bicyclohexyl-ammonium fumagillin, whose trade name was Fumidil-B (Abbot Labs) but now seems to be called Fumagillin-B, is a whitish soluble antibiotic powder discovered in 1952; some beekeepers mix this with sugar syrup and fed to bees to control Nosema disease. Fumagillin is more soluble than Fumidil. Its use in beekeeping is outlawed in the European Union because it is a suspected teratogen (causes birth defects). Fumagillin can block blood vessel formation by binding to an enzyme called methionine aminopeptidase. Targeted gene disruption of methionine aminopeptidase 2 results in an embryonic gastrulation defect and endothelial cell growth arrest. It is made from the fungus that causes stonebrood, Aspergillus fumigatus. Formula: (2E,4E,6E,8E)–10-{[(3S,4S,5S, 6R)-5–methoxy-4-[2–methyl–3-(3–methylbut–2-enyl) oxiran–2-yl]-1-oxaspiro[2.5]octan-6-yl]oxy}-10-oxo-deca-2,4,6,8-tetraenoic acid

Fumidil-B = The old trade name for Fumagillin, see above entry.

Fume board = A device used to hold a set amount of a volatile chemical (A bee repellent like Bee Go or Honey Robber or Bee Quick) to drive bees from supers.

G

Gloves = Leather, cloth or rubber gloves worn while inspecting bees.

Glucose = Also known as dextrose, it is a simple sugar (or monosaccharide) and is one of the two main sugars found in honey; forms most of the solid phase in granulated honey.

Grafting = Removing a worker larva from its cell and placing it in an artificial queen cup in order to have it reared into a queen.

Grafting tool = A needle or probe used for transferring larvae in grafting of queen cells

Granulate = The process by which honey, a super-saturated solution (more solids than liquid) will become solid or crystallize; speed of granulation depends of the kinds of sugars in the honey, the crystal seeds (such as pollen or sugar crystals) and the temperature. Optimum temperature for granulation is 57° F (14° C).

Guard bees = Worker bees about three weeks old, which have their maximum amount of alarm pheromone and venom; they challenge all incoming bees and other intruders.

Gum = A hollow log beehive, sometimes called a log-gum, made by cutting out that portion of a tree containing bees and moving it to the apiary, or by cutting a hollow portion of a log, putting a board on for a lid and hiving a swarm in it. Since it contains no moveable combs, and since each individual state in the US has laws that require movable combs, it is therefore illegal in the US.

H

Hair clip queen catcher = A device used to catch a queen that resembles a hair clip. Available from most beekeeping supply houses.

Haploid = Possessing a single set of genes, as drones do, as opposed to pairs of genes as workers and queens have.

Hemolymph = The scientific name for insect "blood."

Hive = A home for a colony of bees.

Hive body = A wooden box containing frames. Usually referring to the size of box being used for brood.

Hive stand = A structure serving as a base support for a beehive; it helps in extending the life of the bottom board by keeping it off damp ground. Hive stands may be built from treated lumber, cedar, bricks, concrete blocks etc.

Hive staples = Large C-shaped metal nails, hammered into the wooden hive parts to secure bottom to supers, and supers to super before moving a colony.

Hive tool = A flat metal device used to pry boxes and frames apart, typically with a curved scraping surface or a lifting hook at one end and a flat blade at the other.

Hoffman frame = Frames that have the end bars wider than the top bars to provide the proper spacing when frames are placed in the hive. In other words, self spacing frames. In other words, standard frames.

Honey = A sweet viscous material produced by bees from the nectar of flowers, composed largely of a mixture of dextrose and levulose dissolved in about 19 to 17 percent water; contains small amounts of sucrose, mineral matter, vitamins, proteins, and enzymes.

Honey bound = A condition where the brood nest of a hive is being backfilled with honey. This is a normal condition that is used by the workers to shut down the queen's brood production. It usually happens just before swarming and in the fall to prepare for winter.

Honeydew = An excreted material from insects in the order Homoptera (aphids) which feed on plant sap; since it contains almost 90% sugar, it is collected by bees and stored as honeydew honey.

Honey bee = The common name for Apis mellifera.

Honey Bee Healthy = A mixture of essential oils (lemon grass and peppermint) sold to boost the immune system of the bees.

Honey crop = The honey that was harvested.

Honey crop also called honey stomach or honey sac = An enlargement at the posterior of a bees' esophagus but lying in the front part of the abdomen, capable of expanding when full of liquid such as nectar or water. Used for transportation purposes for water, nectar and honey.

Honey extractor = A machine which removes honey from the cells of comb by centrifugal force. The two main types are tangential where the frames lie flat and are flipped to extract the other side, and radial where the frames are like spokes in a wheel and both sides are emptied at the same time.

Honey flow = A time when enough nectar-bearing plants are blooming such that bees can store a surplus of honey.

Honey gate = A faucet used for removing honey from tanks and other storage receptacles.

Honey house = A building used for activities such as honey extraction, packaging and storage.

Honey plants = Plants whose flower (or other parts) yields enough nectar to produce a surplus of honey; examples are asters, basswood, citrus, eucalyptus, goldenrod and tupelo.

Honey Super Cell = Fully drawn plastic comb in deep depth and 4.9mm cell size

Honey supers = Refers to boxes of frames used for honey production. From the Latin "super" for above as a designation for any box above the brood nest.

Hopkins method = A graftless method of queen rearing that involves putting a frame of young larvae horizontally above a brood nest.

Hopkins shim = A shim used to turn a frame flatways for queen rearing without grafting.

Horizontal hive = a hive that is laid out horizontally instead of vertically in order to eliminate lifting boxes.

Hornets and Yellow Jackets = Social insects belonging to the family Vespidae. Nest in paper or foliage material, with only an overwintering queen. Fairly aggressive, and carnivorous, but generally beneficial, they can be a nuisance to man. Hornets and Yellow Jackets are often confused with Wasps and Honey Bees. Wasps are related to Hornets and Yellow Jackets, the most common of which are the paper wasps which nest in small exposed paper combs, suspended by a single support. Hornets, Yellow Jackets and Wasps are easy to distinguish by their shiny hairless body, and aggressiveness. Yellow jackets, unfortunately, look like the bees in the cartoons and advertisements, bright yellow and black and shiny. Honey Bees are generally fuzzy black, brown or tan, never bright yellow, and basically docile in nature.

Hot (temperament) = Bees that are overly defensive or outright aggressive.

Housel positioning theory = A theory proposed by Michael Housel that natural brood nests have a predictable orientation of the "Y" in the bottom of the cells. Basically that when looking at one side an upside down "Y" will appear in the bottom and from the other side a right side up "Y" will appear and the center comb will have a sideways "Y" that is the same from both sides. Basically if we assume a third bar in my notation to make these "Y"s and assume a nine frame hive and each pair is what the comb looks like from that side: ^v ^v ^v ^v >> v^ v^ v^ v^

Hydroxymethyl furfural = A naturally occurring compound in honey that rises over time and rises when honey is heated.

Hypopharyngeal gland = A gland located in the head of a worker bee that secretes "royal jelly". This rich blend of proteins and vitamins is fed to all bee larvae for the first three days of their lives and queens during their entire development.

I

Israeli Acute Paralysis Virus aka IAPV = The virus currently being blamed for CCD. First discovered in Israel where it was quite devastating to colonies.

Illinois = A box that is $6^5/_8$" in depth and the frames are $6^1/_4$" in depth. AKA Medium AKA Western AKA $^3/_4$ depth.

Imirie shim = A device credited to the late George Imirie that is a $^3/_4$" shim with an entrance built in. It allows you to add an entrance between any two pieces of equipment on the hive.

Increase = To add to the number of colonies, usually by dividing those on hand. See Split.

Infertile = Incapable of producing a fertilized egg, as a laying worker or drone laying queen. Unfertilized eggs develop into drones.

Inhibine = Antibacterial effect of honey caused by enzymes and an accumulation of hydrogen peroxide, a result of the chemistry of honey.

Inner cover = An insulating cover fitting on top of the top super but underneath the outer cover, typically with an oblong hole in the center. Used to be called a "quilt board". In the old days these were often made of cloth.

Instar = Stages of larval development. A Honey Bee goes through five instars. The best queens are grafted in the 1st (preferably) or 2nd instar and not later than that.

Instrumental insemination aka II or AI = The introduction of drone spermatozoa into the spermatheca of a virgin queen by means of special instruments

Invertase = An enzyme in honey, which splits the sucrose molecule (a disaccharide) into its two components dextrose and levulose (monosaccharides). This is produced by the bees and put into the nectar to convert it in the process of making honey.

Isomerase = A bacterial enzyme used to convert glucose in corn syrup into fructose, which is a sweeter sugar; called isomerose, is now used as a bee feed.

Italian bees = A common race of bees, Apis mellifera ligustica, with brown and yellow bands, from Italy; usually gentle and productive, but tend to rob and brood incessantly.

J

Jenter = A particular brand of graftless queen rearing system.

K

Kashmir Bee Virus = A widespread disease of bees, spread more quickly by Varroa, found everywhere there are bees.

Kenya Top Bar Hive = A top bar hive with sloped sides. The theory is that they will have less attachments on the sides because of the slope.

Kidneys = Bees don't actually have kidneys. They have malpighian tubules which are thin filamentous projects from the junction of the mid and hind gut of the bee that cleanse the hemolymph (blood) of nitrogenous cell wastes and deposit them as non-toxic uric acid crystals into the undigestible food wastes for elimination. They serve the same purpose in bees as kidneys do in higher animals.

L

Landing board = An extraneous construction that makes small platform at the entrance of the hive for the bees to land on before entering the hive. Usually just a longer bottom board. Sometimes a sloped approach is added. Bees in nature have none. I call it a "mouse ramp" as the only actual purpose I see it provide is a place for mice to get into the hive more conveniently.

Lang = Short for Langstroth hive.

Langstroth, Rev. L.L. = A Philadelphia native and minister (1810-95), he lived for a time in Ohio where he continued his studies and writing of bees; recognized the importance of the bee space, resulting in the development of the most commonly used movable-frame hive.

Langstroth hive = The basic hive design of L.L. Langstroth. In modern terms any hive that takes frames that have a 19" top bar and fit into a box $19^7/_8$" long. Widths vary from five frame nucs to eight frame boxes to ten frame boxes and from Dadant deeps, Langstroth deeps, Mediums, Shallows and Extra Shallow. But all would still be Langstroths. This would distinguish them from WBC, Smith, National DE etc.

Large Cell = Standard foundation size = 5.4mm cell size

Larva, open = The second developmental stage of a bee, starting the 4th day from when the egg is laid until it's capped on about the 9th or 10th day.

Larva, capped = The second developmental stage of a bee, ready to pupate or spin its cocoon (about the 10th day from the egg).

Laying workers = Worker bees which lay eggs in a colony caused by them being a few weeks without the pheromones from open brood; such eggs are infertile, since the workers cannot mate, and therefore become drones.

Leg baskets = Also called pollen baskets, a flattened depression surrounded by curved spines located on the outside of the tibiae of the bees' hind legs and adapted for carrying flower pollen and propolis.

Lemon Grass essential oil = Essential oil used for swarm lure which contains many of the constituents of Nasonov pheromone.

Levulose = Also called fructose (fruit sugar), a monosaccharide commonly found in honey that is slow to granulate.

Long hive = a hive that is laid out horizontally instead of vertically.

M

Malpighian tubules = Thin filamentous projects from the junction of the mid and hind gut of the bee that cleanse the hemolymph of nitrogenous cell wastes and deposit them as non-toxic uric acid crystals into the undigestible food wastes for elimination. They serve the same purpose as kidneys in higher animals.

Mandibles = The jaws of an insect; used by bees to form the honey comb and scrape pollen, in fighting and picking up hive debris.

Marking = Painting a small dot of enamel on the back of the thorax of a queen to make her easier to identify and so you can tell her age and if she has been superseded.

Marking pen = An enamel pen used to mark queens. Available at local hardware stores as enamel pens. Also from beekeeping supply houses as Queen marking pens.

Marking Tube = A plastic tube commonly available from beekeeping supply houses that is used to safely confine a queen while you mark her.

Mating flight = The flight taken by a virgin queen while she mates in the air with several drones.

Mating nuc = A small nuc for the purpose of getting queens mated used in queen rearing. These vary from two frames of the standard size used by that beekeeper for brood, to the mini-mating nucs sold for that purpose with smaller than normal frames. The concept of all mating nucs is to use less resources to get queens mated

Maxant = A beekeeping equipment manufacturer that makes uncappers, extractors, hive tools etc.

Medium = A box that is $6^5/_8$" in depth and the frames are $6^1/_4$" in depth. AKA Illinois AKA Western AKA $^3/_4$ depth.

Medium brood (foundation) = When used to refer to foundation, medium refers to the thickness of the wax *not* the depth of the frame. In this case it's medium thick and of worker sized cells.

Melissococcus pluton = New name given by taxonomists for the bacterium that causes European Foulbrood. The old name was Streptococcus pluton.

Midnite = An F1 hybrid cross of two specific lines of Caucasians and Carniolans. Originated by Dadant and Sons and sold for years by York. Originally they were two lines of Caucasians, but eventually became a cross between Caucasians and Carniolans.

Migratory beekeeping = The moving of colonies of bees from one locality to another during a single season to take advantage of two or more honey flows or for pollination.

Migratory cover = An outer cover used without an inner cover that does not telescope over the sides of the hive; used by commercial beekeepers who frequently move hives. This allows hives to be packed tightly against one another because the cover does not protrude over the sides.

Miller Bee Supply = A beekeeping supply company out of North Carolina. Among other things, they have eight frame equipment.

Miller feeder = Top feeder popularized by C.C. Miller.

Miller Method = A graftless method of queen rearing that involves a ragged edge on some brood comb for the bees to build queen cells on.

Moisture content = In honey, the percentage of water should be no more than 18.6; any percentage higher than that will allow honey to ferment.

Mouse guard = A device to reduce the entrance to a hive so that mice cannot enter. Commonly #4 hardware cloth.

Movable combs = Combs that are built in a hive that allows them to be manipulated and inspected individually. Top bar hives have movable combs but not frames. Langstroth hives have movable combs *in* frames.

Movable frames = A frame constructed in such a way to preserve the bee space, so they can be easily removed; when in place, it remains unattached to its surroundings.

N

Nadiring = Adding boxes below the brood nest. This is a common practice with foundationless including Warre' hives.

Nasonov = A pheromone given off by a gland under the tip of the abdomen of workers that serves primarily as an orientation pheromone. It is essential to swarming behavior and nasonoving is set off by disturbance of the colony. It is a mixture of seven terpenoids, the majority of which is Geranial and Neral, which are a pair of isomers usually mixed and called citral. Lemongrass (Cymbopogon) essential oil is mostly these scents and is useful in bait hives and to get newly hived bees or swarms to stay in a hive.

Nasonoving = Bees who have their abdomens extended and are fanning the Nasonov pheromone. The smell is lemony

Natural cell = Cell size that bees have built on their own without foundation.

Natural comb = Comb that bees have built on their own without foundation.

Nectar = A liquid rich in sugars, manufactured by plants and secreted by nectary glands in or near flowers; the raw material for honey.

Nectar flow = A period of time when nectar is available.

Nectar Management aka Checkerboarding = a method of swarm control originated by Walt Wright where the stores above the brood chamber are alternated with drawn comb late in the winter. Reports from those using it are of massive harvests and no swarming.

New World Carniolans = A breeding program originated by Sue Cobey to find and breed bees from the US with Carniolan traits and other commercially useful traits.

Newspaper method = A technique to join together two colonies by providing a temporary newspaper barrier. Usually one sheet with a small slit. Usually you make sure both colonies can still fly and ventilate.

Nicot = A particular brand of graftless queen rearing system.

Nosema = Disease caused by a fungus (used to be classified as a protozoan) called Nosema apis. The common chemical solution (which I don't use) was Fumidil which has been recently renamed Fumagillin-B. Feeding honey or syrup is an effective remedy. Symptoms are a white distended gut, dysentery and especially seeing Nosema under a microscope from the gut of a field stripped bee.

Nuc, nuclei, nucleus = A small colony of bees often used in queen rearing or the box in which the small colony of bees resides. The term refers to the fact that the essentials, bees, brood, food, a queen or the means to make one, are there for it to grow into a colony, but it is not a full sized colony.

Nurse bees = Young bees, usually three to ten days old, which feed and take care of developing brood.

O

Observation Hive = A hive made largely of glass or clear plastic to permit observation of bees at work

Open-air Nest = A colony that has built its nest in the open limbs of a tree rather than in the hollow of a tree or a hive.

Open Mesh Floor = A bottom board with screen (usually #8 hardware cloth) for the bottom to allow ventilation and to allow Varroa mites to fall through. In the US this is typically called a Screened Bottom Board.

Outer cover = The last cover that fits over a hive to protect it from rain; the two most common kinds are telescoping and migratory covers.

Outyard = Also called out apiary, it is an apiary kept at some distance from the home or main apiary of a beekeeper.

Ovary = The egg producing part of a plant or animal.

Ovule = An immature female germ cell, which develops into a seed.

Ovariole = Any of several tubules that compose an insect ovary.

Oxytetracycline aka Oxytet = An antibiotic sold under the trade name Terramycin; used to control American and European foulbrood diseases.

P

Package bees = A quantity of adult bees (2 to 5 pounds), with or without a queen, contained in a screened shipping cage.

Parasitic Mite Syndrome aka Bee Parasitic Mite Syndrome = A set of symptoms that are caused by a major infestation of Varroa mites. Symptoms include the presence of Varroa mites, the presence of various brood diseases with symptoms similar to that of foulbroods and sacbrood but with no predominant pathogen, AFB-like symptoms, spotty brood pattern, increased supersedure of queens, bees crawling on the ground, and a low adult bee population.

Parasitic Mites = Varroa and tracheal mites are the mites with economic issues for bees. There are several others that are not known to cause any prob-lems.

Paralysis aka APV aka Acute Paralysis Virus = A viral disease of adult bees which affects their ability to use legs or wings normally.

Parthenogenesis = The development of young from unfertilized eggs laid by virgin females (queen or worker); in bees, such eggs develop into drones.

Para Dichloro Benzene (aka PDB aka Para-moth) = Wax moth treatment for stored combs. A known carcinogen.

PermaComb = Fully drawn plastic comb in medium depth and about 5.0mm equivalent cell size after allowing for cell wall thickness and taper of the cell..

PF-100 (deep) and PF-120 (medium) = A small cell one piece plastic frame available from Mann Lake. Measures 4.95mm cell size. Users report excellent acceptance and perfectly drawn cells.

Phoretic = In the context of Varroa mites it refers to the state where they are on the adult bees instead of in the cell either developing or reproducing.

Piping = A series of sounds made by a queen, frequently before she emerges from her cell. When the queen is still in the cell it sounds sort of like a quack quack quack. When the queen has emerged it sounds more like zoot zoot zoot.

Play flights aka orientation flights = Short flights taken in front and in the vicinity of the hive by young bees to acquaint them with hive location; sometimes mistaken for robbing or swarming preparations.

Pollen = The dust-like male reproductive cells (gametophytes) of flowers, formed in the anthers, and important as a protein source for bees; fermented pollen (bee bread) is essential for bees to rear brood.

Pollen basket = An anatomical structure on the bees legs where pollen and propolis is carried.

Pollen bound = A condition where the brood nest of a hive is being filled with pollen so that there is nowhere for the queen to lay.

Pollen box = A box of brood moved to the bottom of the hive during the honey flow to induce the bees to store pollen there, or a box of pollen frames that was put on the bottom purposefully. This provides pollen stores for the fall and winter. The term was coined by Walt Wright.

Pollen pellets or cakes = The pollen packed in the pollen baskets of bees and transported back to the colony made by rolling in the pollen, brushing it off and mixing it with nectar and packing it into the pollen baskets.

Pollen substitute = A food material which is used to substitute wholly for pollen in the bees' diet; usually contains all or part of soy flour, brewers' yeast, wheast, powdered sugar, or other ingredients. Research has shown that bees raised on substitute are shorter lived than bees raised on real pollen.

Pollen supplement = A mixture of pollen and pollen substitutes used to stimulate brood rearing in periods of pollen shortage

Pollen trap = A device for collecting the pollen pellets from the hind legs of worker bees; usually forces the bees to squeeze through a screen mesh, usually #5 hardware cloth, which scrapes off the pellets which fall through #7 hardware cloth into a drawer with a screened bottom so the pollen won't mold.

Porter bee escape = Introduced in 1891, the escape is a device that allows the bees a one-way exit between two thin and pliable metal bars that yield to the bees' push; used to free honey supers of bees but may clog since drone bees often get stuck.

Prime swarm = The first swarm to leave the parent colony, usually with the old queen.

Proboscis = The mouthparts of the bee that form the sucking tube or tongue

Propolis = Plant resins collected, mixed with enzymes from bee saliva and used to fill in small spaces inside the hive and to coat and sterilize everything in the hive. It has antimicrobial properties. It is typically made from the waxy substance from the buds of the poplar family but in a pinch may be anything from tree sap to road tar.

Propolize = To fill with propolis, or bee glue.

Pupa = The third stage in the development of the bee during which it is inactive and sealed in its cocoon.

Push In Cage = Cage made of #8 hardware cloth used to introduce or confine queens to a small section of comb. Usually used over some emerging brood.

Q

Queen = A fully developed female bee responsible for all the egg laying of a colony.

Queen Bank = Putting multiple caged queens in a nuc or hive.

Queen cage = A special cage in which queens are shipped and/or introduced to a colony, usually with 4 to 7 young workers called attendants, and usually a candy plug.

Queen cage candy = Candy made by kneading powdered sugar with invert sugar syrup until it forms a stiff dough; used as food in queen cages.

Queen cell = A special elongated cell resembling a peanut shell in which the queen is reared; usually over an inch in length, it hangs vertically from the comb.

Queen clipping = Removing a portion of one or both wings of a queen to prevent her from flying or to better identify when she has been replaced.

Queen cup = A cup-shaped cell hanging vertically from the comb, but containing no egg; also made artificially of wax or plastic to raise queens

Queen excluder = A device made of wire, wood or zinc (or any combination thereof) having openings of .163 to .164 inch, which permits workers to pass but excludes queens and drones; used to confine the queen to a specific part of the hive, usually the brood nest.

Queen juice = When retired queens are added to a jar of alcohol, that alcohol becomes "Queen juice". It contains QMP and is good for swarm lure.

Queenright = A colony that contains a queen capable of laying fertile eggs and making appropriate pheromones that satisfy the workers of the hive that all is well.

Queen Mandibular Pheromone aka Queen substance aka QMP = A pheromone produced by the queen and fed to her attendants who share it with the rest of the colony that gives the colony the sense of being queenright. Chemically QMP is very diverse with at least 17 major components and other minor ones. 5 of these compounds are: 9-ox-2-decenoic acid (9ODA) + cis & trans 9 hydroxydec-2-enoic acid (9HDA) + methyl-p-hydroxybenzoate (HOB) and 4-hydroxy-3-methoxyphenylethanol (HVA). Newly emerged queens produce very little of this. By the sixth day they are producing enough to attract drones for mating. A laying queen makes twice that amount. QMP is responsible for inhibition of rearing replacement queens, attraction of drones for mating, stabilizing and organizing a swarm around the queen, attracting a retinue of attendants, stimulating foraging and brood rearing, and the general moral of the colony. Lack of it also seems to attract robber bees.

Queen muff = A screen wire tube that resembles a "muff" to keep your hands warm in shape but is used to keep queens from escaping when marking them or releasing attendants. Available from Brushy Mountain.

R

Rabbet = In wood working a groove cut into wood. The frame rests in a Langstroth hive are rabbets and the corners are sometimes done as rabbets and sometimes as finger or box joints.

Races of Bees = In taxonomy this is actually a variety but in beekeeping it is typically called a "race". All of these are Apis mellifera. The most common currently In the US are Italians (ligustica), Carniolans (carnica) and Caucasians (caucasica). Russians would be either carpatica, acervorum, carnica or caucasica depending on who you are talking to.

Radial extractor = A centrifugal force machine to throw out honey but leave the combs intact; the frames are placed like spokes of a wheel, top bars towards the wall, to take advantage of the upward slope of the cells.

Rauchboy = A particular brand of smoker that has an inner chamber to provide more consistent oxygen to the fire.

Raw honey = Honey that has not been finely filtered or heated.

Regression = As applied to cell size, large bees, from large cells, cannot build natural sized cells. They build something in between. Most will build 5.1 mm worker brood cells. Regression is getting large bees back to smaller bees so they can and will build smaller cells.

Reorientation = When the bees take note of their surroundings and landmarks to make sure they remember the location of the colony. A variety of things set this off. Young bees will orient (not reorient but it's the same behavior) when they first emerge from the hive. A virgin queen will orient for a day or so before going on her nuptials. Confining tends to set this off. Even short confinements will cause some to reorient. Confining for 72 hours causes virtually all of them to reorient. When it warms up and they can fly, they will hover around the hive and reorient. Reorientation is triggered even by lower times but the amount of it maxes out at 72 hours. More time won't make any noticeable difference. Obstructions add to reorientation (leaves in the entrance, a branch in front etc.) as does general disruption such as drumming or knocking the hive around a bit. On a warm day shaking a frame or two of bees back into the hive from the combs tends to set off Nasonoving which also tends to set off reorient-ing

Requeen = To replace an existing queen by re-moving her and introducing a new queen.

Rendering wax = The process of melting combs and cappings and removing refuse from the wax.

Retinue = Worker bees that are attending the queen.

Reversing aka Switching = The act of exchang-ing places of different hive bodies of the same colony;

usually for the purpose of nest expansion, the super full of brood and the queen is placed below an empty super to allow the queen extra laying space.

Robber screen = A screen used to foil robbers but let the local residents into the hive.

Robbing = The act of bees stealing honey/nectar from the other colonies; also applied to bees cleaning out wet supers or cappings left uncovered by beekeepers and sometimes used to describe the beekeeper removing honey from the hive.

Ropy = A quality of forming an elastic rope when drawn out with a stick. Used as a diagnostic test for American foulbrood.

Round sections = Sections of comb honey in plastic round rings instead of square wooden boxes, usually Ross Rounds.

Rolling = A term to describe what happens when a frame is too tight or pulled out too quickly and bees get pushed against the comb next to it and "rolled". This makes bees very angry and is sometimes the cause of a queen being killed.

Royal jelly = A highly nutritious, milky white secretion of the hypopharyngeal gland of nurse bees; used to feed the queen and young larvae.

Russian bees = Apis mellifera acervorum or carpatica or caucasica or carnica. Some even say they are crossed with Apis ceranae (very doubtful). They came from the Primorsky region of Russia. They were used for breeding mite resistance because they were already surviving the mites. They are a bit defensive, but in odd ways. They tend to head butt a lot while not necessarily stinging any more. Any first cross of any race may be vicious and these are no exception. They are watchful guards, but not usually "runny" (tending to run around on the comb where you can't find the queen or work with them well). Swarminess and productivity are a bit more unpredictable. Traits are not well fixed. Frugality is similar to the Carniolans. They were brought to the USA by the USDA in June of 1997, studied on an island in Louisiana and then field testing in other states in 1999. They went on sale to the general public in 2000.

S

Sac Brood Virus = Symptoms are the spotty brood patterns as other brood diseases but the larvae are in a sack with their heads raised.

Sclerite = Same as Tergite. An overlapping plate on the dorsal side of a arthropod that allows it to flex.

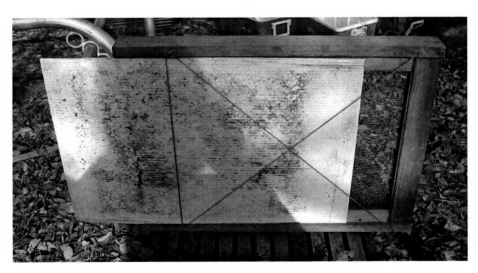

Screened Bottom Board = A bottom board with screen (usually #8 hardware cloth) for the bottom to allow ventilation and to allow Varroa mites to fall through. In Europe this is called an Open Mesh Floor.

Scout bees = Worker bees searching for a new source of pollen, nectar, propolis, water, or a new home for a swarm of bees.

Scutum = Shield shaped portion of the back of the thorax of some insects including Apis mellifera (honey bees). Usually divided into three areas: the anterior prescutum, the scutum, and the smaller posterior scutellum.

Sections = Small wooden (or plastic) boxes used to produce comb honey.

Self-spacing frames aka Hoffman frames = Frames constructed so that everything but the end bar (which is the spacer) is a bee space apart when pushed together in a hive body.

Settling tank = A large capacity container used to settle extracted honey; air bubbles and debris will float to the top, clarifying the honey.

Shallow = A box that is $5^{11}/_{16}$ or $5^3/_4$" deep with frames that are $5^1/_2$" deep.

Shaken swarm = An artificial swarm made by shaking bees off of combs into a screened box and then putting a caged queen in until they accept her. One method for making a divide. Also the method used to make packages of bees.

Skep = A beehive without moveable combs, usually made of twisted straw in the form of a basket; its use is illegal in each state in the U.S as the combs are not inspectable.

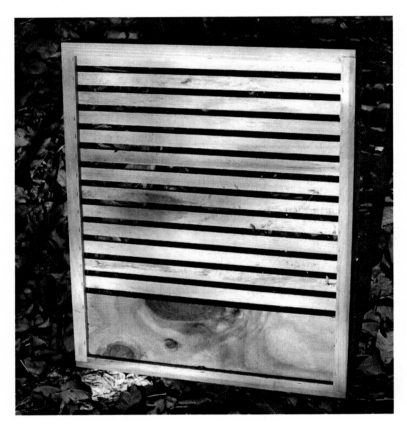

Slatted rack = A wooden rack that fits between the bottom board and hive body. Bees make better use of the lower brood chamber with increased brood rearing, less comb gnawing, and less congestion at the front entrance. Popularized by C.C. Miller and Carl Killion.

Slumgum = The refuse from melted combs and cappings after the wax has been rendered or removed; usually contains cocoons, pollen, bee bodies and dirt.

Small Cell = 4.9mm cell size. Used by some bee-keepers to control Varroa mites.

Small Hive Beetle = A pest recently imported to North America, whose larvae will destroy comb and ferment honey.

Smith method = A method of queen rearing popularized by Jay Smith, that uses a swarm box as a cell starter and grafting larvae into queen cups.

Smoker = A metal container with attached bellows which burns various fuels to generate smoke; used to interfere with the ability to smell alarm pheromone and therefore control aggressive behavior of bees during colony inspections.

Solar wax melter = A glass-covered box used to melt wax from combs and cappings using the heat of the sun.

Sperm cells = The male reproductive cells (gametes) which fertilize eggs; also called spermatozoa.

Spermatheca = A small sac connected with the oviduct of the queen bee in, which is stored, the spermatozoa received by the queen when mating with drones.

Spiracles = Openings into the respiratory system on a bee that can be closed at will. These are on the sides of the bee. They are considerably smaller than the Trachea they protect. The first thoracic spiracle is the one that is infiltrated by the tracheal mites as it is the largest. When closed the spiracles are air tight.

Split = To divide a colony for the purpose of increasing the number of hives.

Spur embedder = A device used for mechanically embedding wires into foundation by employing hand pressure as opposed to using electricity to melt the wires into the wax.

Starline = An Italian bee hybrid known for vigor and honey production. It was an F1 cross of two specific lines of Italian bees. Originated by Dadant and sons and produced for many years by York.

Starter hive aka a Swarm box = A box of shaken bees used to start queen cells.

Sting = An organ belonging exclusively to female insects developed from egg laying mechanisms, used to defend the colony; modified into a piercing shaft through which venom is injected. On workers this has a barb which causes it to catch and pull out.

Streptococcus pluton = Deprecated (old) name for the bacterium that causes European Foulbrood. The new name is Melissococcus pluton.

Sucrose = A polysaccharide. The principal sugar found in nectar. Honey bees break this into Dextrose and Fructose with enzymes.

Sugar syrup = Feed for bees, containing sucrose or table (cane or beet) sugar and hot water in various ratios; usually 1:1 in the spring and 2:1 in the fall.

Sugar roll test = A test for Varroa mites that involves rolling a cupful of bees in powdered sugar and counting the number of mites dislodged. This was invented as a non-lethal alternative to an alcohol wash or an ether roll.

Super = A box with frames in which bees store honey; usually placed above the brood nest. From the Latin *super* meaning "above".

Supering = The act of placing honey supers on a colony in expectation of a honey flow.

Supersedure = Rearing a new queen to replace the mother queen in the same hive; shortly after the daughter queen begins to lay eggs, the mother queen often disappears.

Suppressed Mite Reproduction aka SMR = Queens from a breeding program by Dr. John Harbo that have less Varroa problems probably due to increased hygienic behavior. Lately renamed VSH aka Varroa Sensitive Hygiene.

Surplus (foundation) = Refers to thin foundation used for cut comb honey. The name is referring to the extra sheets of foundation you get from a pound of wax.

Surplus honey = Any extra honey removed by the beekeeper, over and above what the bees require for their own use, such as winter food stores.

Survivor stock = Bees raised from bees that were surviving without treatments. Often feral stock.

Swarm = A temporary collection of bees, containing at least one queen that split apart from the mother colony to establish a new one; a natural method of propagation of honey bee colonies.

Swarm box aka a Starter hive = A box of shaken bees used to start queen cells.

Swarm cell = Queen cells usually found on the bottom of the combs before swarming.

Swarm commitment = The point just after swarm cutoff where the colony is committed to swarming.

Swarm cutoff = The point at which the colony decides to swarm or not. Past this point they either commit to swarming or they commit to just looking out for colony stores for the coming winter.

Swarm trap aka Bait hive aka Decoy hive = A hive placed to attract stray swarms.

Swarm preparation = The sequence of activities of the bees that is leading up to swarming. Visually you can see this start at backfilling the brood nest so that the queen has nowhere to lay.

Swarming = The natural method of propagation of the honey bee colony.

Swarming season = The time of year, usually late spring to early summer, when swarms usually issue.

T

Tanzanian Top Bar Hive = A top bar hive with vertical sides.

Telescopic cover = A cover with a rim that hangs down all the way around it usually used with a inner cover under it.

Ten frame = A box made to take ten frames. $16^{1}/_{4}$" wide.

Terramycin = Called oxytet in Canada and other locations. It is an antibiotic that is often used as a preventative for American and a cure for European foulbrood diseases.

Tested queen = A queen whose progeny shows she has mated with a drone of her own race and has other qualities which would make her a good colony mother. One that has been given time to prove what her qualities are.

Tergal = Pertaining to the Tergum.

Tergite = A hard overlapping plate on the dorsal portion of an arthropod that allows it to flex. Also known as sclerite.

Tergum (plural terga) = The dorsal portion of an arthropod.

Thelytoky = A type of parthenogenetic reproduction where unfertilized eggs develop into females. Usually with bees this is referring to a colony rearing a queen from a laying worker egg. This is very rare, but documented, with European Honey Bees. It is common with Cape Bees.

Thin surplus foundation = A comb foundation used for comb honey or chunk honey production which is thinner than that used for brood rearing. Thinner than surplus.

Thorax = The central region of an insect to which the wings and legs are attached.

Tiger striped (queen) = Markings of a particular type on a queen. Not striped like a worker (who have very even bands) but more like "flames".

Top bar = The top part of a frame or, in a top bar hive, just the piece of wood from which the comb hangs.

Top Bar Hive = a hive with only top bars and no frames that allows for movable comb without as much carpentry or expense.

Top feeder = Miller feeder. A box that goes on top of the hive that contains the syrup. See Miller Feeder.

Top supering = The act of placing honey supers on *top* of the top super of a colony as opposed to putting it under all the other supers, and directly on top of the brood box, which would be *bottom* supering or adding boxes below the brood box which would be nadiring.

Tracheal Mites = A mite that infests the trachea of the honey bee. Resistance to tracheal mites is easily bred for.

Transferring or cut out = The process of changing bees and combs from trees, houses or bee gums or skeps to movable frame hives.

Travel stains = The darkened appearance on the surface of honeycomb caused by bees walking over its surface.

Triple-wide = A box that is three times as wide as a standard ten frame box. $48^3/_4$".

Trophallaxis = The transfer of food or pheromones among members of the colony through mouth-to-mouth feeding. It is used to keep a cluster of bees alive as the edges of the cluster collect food and share it through the cluster. It is also used for communication as pheromones are shared. One very important one is QMP (Queen Mandibular Pheromone) which is shared by trophallaxis throughout the hive.

Twelve frame = A box made to take twelve frames. This is $19^7/_8''$ by $19^7/_8''$.

Two Queen Hive = A management method where more than one queen exists in a hive. The purpose is you get more bees and more honey with two queens.

U

Uncapping knife = A knife used to shave off the cappings of sealed honey prior to extraction; hot water, steam or electricity can heat the knives.

Uncapping tank = A container over which frames of honey are uncapped; usually strains out the honey which is then collected.

Unfertilized = An ovum or egg, which has not been united with the sperm.

Uniting = Combining two or more colonies to form a larger colony. Usually done with a sheet of newspaper between.

Unlimited Brood Nest aka "food chamber" = running bees in a configuration where the brood nest is not limited by an excluder and they are usually overwintered in more boxes to allow more food and more expansion in the spring.

V

Varroa destructor used to be called Varroa Jacobsoni = Parasitic mite of the honey bee.

Veil = A protective netting or screen that covers the face and neck; allows ventilation, easy movement and good vision while protecting the primary targets of guard bees.

Venom allergy = A condition in which a person, when stung, may experience a variety of symptoms ranging from hives to anaphylactic shock. A person who is stung and experiences systemic (the whole body or places remote from the sting) symptoms should consult a physician before working bees again.

Venom hypersensitivity = A condition in which a person, if stung, is likely to experience an anaphylactic shock. A person with this condition should carry an emergency insect sting kit at all times during warm weather

Virgin queen = An unmated queen bee.

W

Walter T. Kelley = A beekeeping supply company out of Clarkson, KY. They have many things no one else does.

Warré hive = A type of vertical top bar hive invented by Abbé Émile Warré.

Washboarding = When the bees on the landing board or the front of a hive are moving in unison resembling a line dance.

Warming cabinet = An insulated box or room heated to liquefy honey or to heat honey to speed extraction.

Wax Dipping Hives = A method of protecting wood and also of sterilizing from AFB where the equipment is "fried" in a mixture of wax and gum resin. Usually done with paraffin sometimes done with beeswax.

Wax glands = The eight glands located on the last 4 visible, ventral abdominal segments of young worker bees; they secrete beeswax flakes.

Wax moths = See chapter *Enemies of the Bees*. Wax moths are opportunists. They take advantage of a weak hive and live on pollen, honey and burrow through the wax.

Wax scale or flake = A drop of liquid beeswax that hardens into a scale upon contact with air; in this form it is shaped into comb.

Wax tube fastener = A metal tube for applying a fine stream of melted wax to secure a sheet of foundation into a groove on a frame.

Western = A box that is $6^5/_8$" in depth and the frames are $6^1/_4$" in depth. AKA Illinois AKA Medium AKA $^3/_4$ depth.

Western Bee Supply = A beekeeping supply company out of Montana. The company that makes all of Dadant's equipment. Also sell eight frame equipment.

Windbreaks = Specially constructed, or naturally occurring barriers to reduce the force of the (winter) winds on a beehive.

Winter cluster = A tight ball of bees within the hive to generate heat; forms when outside temperature falls below 50º F.

Winter hardiness = The ability of some strains of honeybees to survive long winters by frugal use of stored honey.

Wire, frame = Thin 28# wire used to reinforce foundation destined for the broodnest or honey extractor.

Wire cone escape = A one-way cone formed by window screen mesh used to direct bees from a house or tree into a temporary hive.

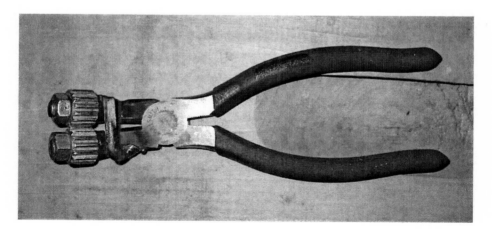

Wire crimpers = A device used to put a ripple in the frame wire to both make it tight and to distribute stress better and give more surface to bind it to the wax.

Worker bees = Infertile female bee whose reproductive organs are only partially developed, and is anatomically different than a queen and is equipped and responsible for carrying out all the routine duties of the colony.

Worker comb = Comb measuring between 4.4mm and 5.4mm, in which workers are reared and honey and pollen are stored.

Worker Queen aka laying workers = Worker bees which lay eggs in a colony hopelessly queenless; such eggs are not fertilized, since the workers cannot mate, and therefore become drones.

Worker policing = Workers that remove eggs laid by workers.

Y

Yellow (queen or bees) = When used to refer to honey bees this refers to a lighter brown color. Honey bees are *not* yellow. A Yellow queen is usually a solid light brown.

Appendix to Volume I: Acronyms

ABJ = American Bee Journal. One of the two main bee magazines in the USA.

AFB = American Foulbrood

AHB = Africanized Honey Bees

AM = Apis mellifera. (European honey bees)

AMM = Apis mellifera mellifera

APV = Acute Paralysis Virus. This virus kills both adult bees and brood.

BC = Bee Culture aka Gleanings in Bee Culture. One of the two main Beekeeping magazines in the USA

BLUF = Bottom Line Up Front. A style of writing where you present the conclusion at the beginning. Common in scientific studies or military correspondence.

BPMS = Bee Parasitic Mite Syndrome

Carni = Carniolan = Apis mellifera carnica

Cauc = Caucasian = Apis mellifera Caucasia

CB = Checkerboarding (aka Nectar Management)

CCD = Colony Collapse Disorder

CPV = Chronic Paralysis Virus

CW = Conventional Wisdom

DCA = Drone Congregation Area

DVAV = Dorsal-Ventral Abdominal Vibrations dance.

DWV = Deformed Wing Virus

EAS = Eastern Apiculture Society

EFB = European Foulbrood

EHB = European Honey Bees

FGMO = Food Grade Mineral Oil

FWIW = For What It's Worth.

FWOF = Floor With Out a Floor

HAS = Heartland Apiculture Society

HBH = Honey Bee Healthy

HBTM = Honey Bee Tracheal Mite

HFCS = High Fructose Corn Syrup. A common bee feed.

HSC = Honey Super Cell (Fully drawn plastic comb in deep depth and 4.9mm cell size)

HMF = Hydroxymethyl furfural. A naturally occurring compound in honey that rises over time and rises when honey is heated.

IAPV = Israeli Acute Paralysis Virus. The virus currently being blamed for CCD

IPM = Integrated Pest Management

IMHO = In My Humble Opinion

IMO = In My Opinion

IMPOV = In My Point Of View

KTBH = Kenya Top Bar Hive (one with sloped sides)

KBV = Kashmir Bee Virus

LC = Large Cell (5.4mm cell size)

LGO = Lemon Grass (essential) Oil (used for swarm lure)

MAAREC = Mid-Atlantic Apiculture Research and Extension Consortium

NM = Nectar Management (aka Checkerboarding)

NWC = New World Carniolans

OA = Oxalic Acid. An organic acid used to kill Varroa as either a syrup or vaporized.

OSR = Oil Seed Rape (aka Canola). A crop that produces honey that is grown to produce oil.

PC = PermaComb (Fully drawn plastic comb in medium depth and about 5.0mm cell size)

PDB = Para Dichloro Benzene (aka Paramoth wax moth treatment)

PMS = Parasitic Mite Syndrome

QMP = Queen Mandibular Pheromone

SBB = Screened Bottom Board

SBV = Sac Brood Virus

SC = Small Cell (4.9mm cell size)

SHB = Small Hive Beetle

SMR = Suppressed Mite Reproduction (usually referring to a queen)

TBH = Top Bar Hive

TM = Terramycin or Tracheal Mites depending on the context

T-Mites = Tracheal Mites

TTBH = Tanzanian Top Bar Hive (one with vertical sides)

ULBN = Unlimited Brood Nest

VD = Varroa destructor

VJ = Varroa jacobsoni

V-Mites = Varroa Mites

VSH = Varroa Sensitive Hygiene. Similar to and appears to be a more specific name for the SMR trait. A trait in queens that is being bred for where the workers sense Varroa infested cells and clean them out.

About the Author

"His writing is like his talks, with more content, detail, and depth than one would think possible with such few words...his website and PowerPoint presentations are the gold standard for diverse and common sense beekeeping practices."—Dean Stiglitz

Michael Bush has had an eclectic set of careers from printing and graphic arts, to construction to computer programming and a few more in between. Currently he is working in computers. He has been keeping bees since the mid 70's, usually from two to seven hives up until the year 2000. Varroa forced more experimentation which required more hives and the number has grown steadily over the years from then. By 2008 it was about 200 hives. He is active on many of the Beekeeping forums with last count at about 45,000 posts between all of them. He has a web site on beekeeping at www.bushfarms.com/bees.htm